TRUST ME:

AI GOVERNANCE

Greg Hutchins PE CERM

Margaux Hutchins CERM

Special thanks to our book reviewers:

Margaux Hutchins

Carolyn Turbyfill

Ed May

Henry Lindborg

James Kline Ph.D.

Dan Swanson

Table of Contents

CHAPTER 0
PREFACE

What is the Key Idea in this Chapter?
The key idea in the Preface is AI is the most important technology of our lifetimes, and we need to understand it.

AI: Most Important Technology of Our Lifetime

AI impacts lots of humans. One mistake can affect millions of humans Think of the consequences if an AI system had access to nuclear launch codes. Or, if a glitch in AI trading crashes stock markets in seconds.

Many compare AI to the birth of the Industrial Age. And others say AI is the electricity of today. Marc Benioff, Salesforce CEO, distilled the importance of AI:

> "Artificial intelligence and generative AI may be the most important technology of our lifetime."[1]

Trust Me Series
Trust Me: AI Governance explains how to understand and work with AI. This book helps you develop personal tools to engage with this disruptive technology. We call these tools for humans to work with AI as 'rules of engagement' or simply AI governance.

So, what is trustworthy AI? Jenny Lyons – Cunha, a Built-in writer defines it as:

"Trustworthy AI refers to artificial intelligence systems that are designed to ensure reliability, fairness, transparency and accountability, while also prioritizing ethical considerations and human rights. These systems operate within legal and regulatory frameworks, respect privacy, mitigate biases and are protected against adversarial attacks or misuse."[2]

Your Use Cases: How To Use This Book

The following are personal use cases for **Trust Me: AI Governance**:

- Use **Trust Me: AI Governance** as your personal AI tutor. *Fortune Magazine* cover shouted: "Nvidia CEO Jensen Huang Says Everyone Should Get an AI Tutor Right Away."
- Use **Trust Me: AI Governance** as your roadmap to get the #1 future job - AI Ethics Steward. McKinsey, the strategy consulting firm, says: "AI Ethics Stewards (who) monitor AI systems development and use" is the #1 new role brought about by Gen AI adoption."[3]
- Over the next few years, AI will disrupt all work, professions, and careers. Autonomous decision-making will be everywhere. Humans-the-middle will have to arbitrate autonomous decisions, so they are fair, transparent, and trustworthy for humans. Picking up on the above bullet, AI Ethics Stewards will be autonomous decision arbiters.
- Implement AI governance to establish trust. Trust or more specifically the lack of it is the #1 inhibitor to AI adoption.
- Understand, comply and manage EU AI Act high-risk AI systems.
- Comply with EU AI Act Literacy requirement using this book as a textbook, training manual, or procedural manual.
- Provide basic knowledge to help humans upskill into the world of data and AI.

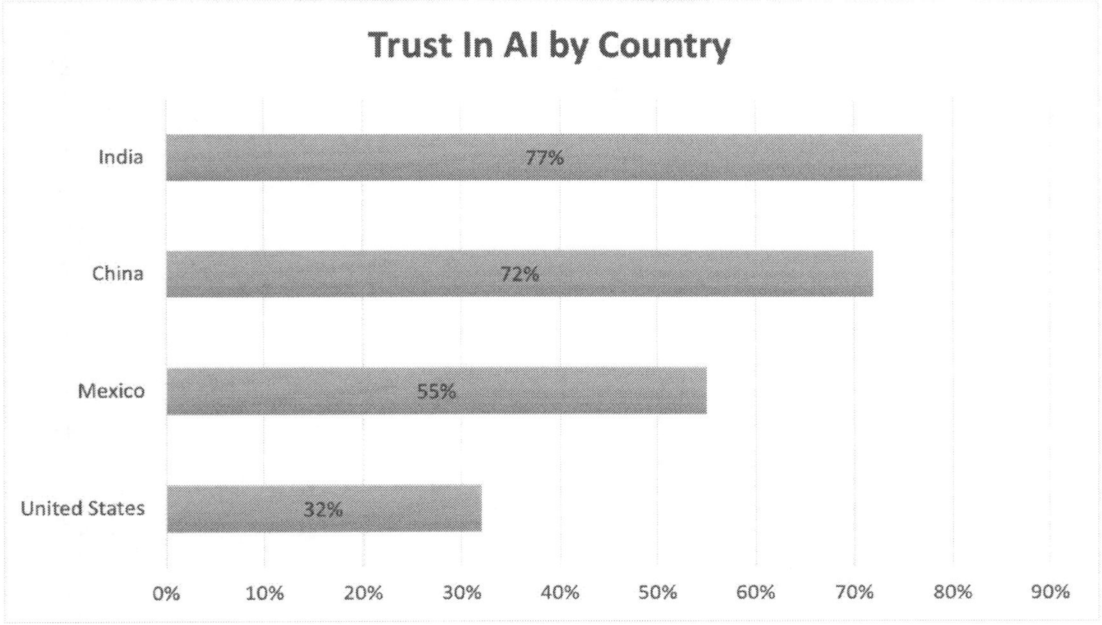

- Pass IAPP's Certified AI Governance Professional certification.
- Comply with AI global regulations like GDPR and OECD Principles.
- Learn ISO 31000 Enterprise Risk Management, which is an EU AI Act requirement.
- Learn and understand the difference between AI compliance and AI assurance of high-risk, AI systems.
- Tailor risk assessment and management frameworks to identify and mitigate AI risks.
- Assure human-in-the-loop processes for critical decision-making controls.
- Understand terms and concepts in AI governance, risk management, and compliance.
- Save hundreds using open-source AI standards and resources. For example, ISO 42001 standard costs $225 in the U.S.

Trust Me: AI Governance Chapters

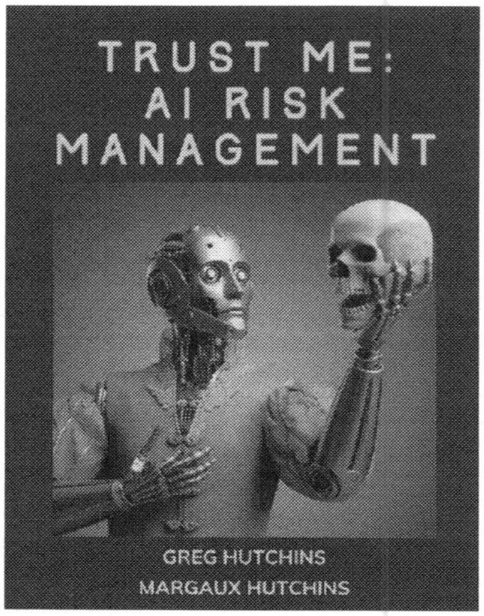

This is the third book in the Trust Me series of books. The first book was **Trust Me: AI Risk Management**. The second book was **Trust Me: ISO 42001**. The third book is **Trust Me: AI Governance**.

Why did we develop the Trust Me series? Well, most of the world does not trust AI. You can see the trust numbers in the figure on the previous page. In the U.S., only 32% of the people trust AI.

Trust Me: AI Governance Organization

Each chapter in **Trust Me: AI Governance** discusses a critical element of the human and machine interaction or in our words, human and machine engagement.

Trust Me: AI Governance covers the following:

- **Chapter 0 - Preface:** The key idea in the Preface is AI is the most important technology of our lifetimes and we need to understand it.
- **Chapter 1 - What is AI?:** The key idea in this chapter is AI is evolving from a deep technical discipline into a socio-technical discipline.
- **Chapter 2 - What is AI Governance?:** The key idea in this chapter is that AI is everywhere. This implies AI governance will soon be everywhere.
- **Chapter 3 - AI Rules of Engagement:** The key idea in this chapter is AI guard rails are the essence of AI governance and the basis for human and machine rules of engagement.
- **Chapter 4 - AI Privacy:** The key idea in this chapter is personal information should stay personal and private.

- **Chapter 5 - EU AI Act:** The key idea in this chapter is the EU has taken the global lead in AI regulation and governance.
- **Chapter 6 - AI Standards & Guidelines:** The key idea in this chapter is understanding AI standards and guidelines is critical for AI governance.
- **Chapter 7 - ISO 31000 Enterprise Risk Management:** The key idea in this chapter is ISO 31000 will become the ERM solution for AI governance of high-risk AI systems.
- **Chapter 8 - Architecting AI Systems:** The key idea in this chapter is AI governance must be architected and tailored into AI systems to identify and mitigate risks and biases inherent in these systems.
- **Chapter 9 - Designing AI Systems:** The key idea in this chapter is understanding how AI systems are designed and built is critical for effective AI governance.
- **Chapter 10 - Deploying AI Systems:** The key idea in this chapter is understanding the critical steps of moving an AI system into a real-world environment helps to develop realistic AI governance.
- **Chapter 11 - Assuring AI Systems:** The key idea in this chapter is assurance, monitoring, and testing are key to building trust in AI systems. AI assurance is the essence of AI trust.
- **Chapter 12 - AI Futures:** The key idea in this chapter is AI is coming quickly, will require governance, will generate opportunities, and will be a bumpy ride.

Lesson Learned - Something to Think About?

We normally think AI is a workplace tool. Yes. It is. For example, Anthropic's chatbot Claude is used for app development (10.4%) content creation and communication, and academic research (7.2%).[4] Chatbots are used as avatar friending, personal stylists, cooking companions, and real time translators. So, how are you using AI systems?

AI Stories

AI is a tough topic. AI governance books seem to be apocalyptic. The purpose of **Trust Me: AI Governance** is to make AI governance understandable, explainable, and trustworthy by taking a broad look at AI legislation, regulations, practices, and standards.

Trust Me: AI Governance stories will help you develop AI strategies and tactics. You will learn what AI can and cannot do; when to use AI; when to ask for help; and how to spot problems and risks.

AI Governance Quotes

You will find lots of quotations in **Trust Me: AI Governance**. They are from the leading thinkers and doers in AI governance. We want you to get different perspectives and arguments.

There are AI doomer and boomer quotations. Depending on the context, we waffle between the two positions. However both positions make sense especially as the technology grows exponentially and trust becomes a huge topic. As well, the quotes help to highlight special topics and points of view.

Our AI Development Stories

In **Trust Me: AI Governance**, we do something different. We share our AI development stories of building the Artificial Intelligence Risk Management Framework (AI RMF) App® including chatbots, RAGs, agents, and AI systems.

In **Trust Me: AI Governance**, each section has a *Lesson Learned* from a personal story addressing AI governance and AI risk management.

In **Trust Me: AI Governance**, we take a systems view of AI governance. We look at AI governance as risk frameworks, policies, and processes to develop AI systems and AI models. Based on our experiences, we follow the Architect, Design, Deploy and Assure® framework throughout this book.

AI Development and Governance Stories

Throughout **Trust Me: AI Governance**, we share AI development governance stories because they:

- Show how AI governance principles play out in real world contexts.
- Demonstrate 'rules of engagement' ideas that work in real life and practice.
- Provide concrete examples to understand and internalize abstract AI governance concepts and complex ideas.
- Share how providers/developers look at AI governance.
- Provide stories of developers and engineers making risk-based decisions and AI governance tradeoffs.
- Help readers understand AI governance concepts and how to apply them.

Quality Plus Engineering

We are a risk management and risk assurance company in the U.S. that does Critical Infrastructure Protection: Forensics, Assurance, Analytics®. Why it matters? There are 16 critical infrastructures in the U.S. Each has critical autonomous decision-making. The EU is regulating these high-risk AI systems.

AI technologies, neural networks, and algorithms have been around for 50 or so years. For example, we were involved with AI starting with LISP language 30 years ago. We have designed neural nets and algorithms.

Use of Terms

Note: In this book, we use the terms 'AI system,' 'AI model,' 'AI agent,' and 'AI app' differently. 'AI app' is a software application that uses AI techniques to provide end users with specific information or customer support.

AI model is trained and tested algorithm or program to do specific things such as pattern recognition and simple decision-making. Usually, AI model is a process with input data and output results.

On the other hand, 'AI system' is a broader machine-based system that can make autonomous risk-based decisions. AI system is more complex and can integrate multiple AI models. The AI system can also learn and adapt based on inputs and a process to do specific things on its own.

Trust Me: AI Governance will be updated yearly. So, if you have feedback such as opportunities for improvement or something we missed, we will correct our errors in the new edition. Thank you for purchasing this book. We look forward to hearing your thoughts.

Greg Hutchins PE CERM – Greg.B.Hutchins@gmail.com
Margaux Hutchins CERM – MargauxHutchins@gmail.com

CHAPTER 1

WHAT IS AI?

What is the Key Idea in this Chapter?

The key idea in this chapter is AI is evolving from a deep technical discipline into a socio-technical discipline. Why? AI is everywhere. AI autonomous decision-making socially impacts all of us. And AI is here to stay.

This is a book about AI governance. One of the first things AI professionals need to do is develop a governance, risk, and compliance taxonomy in their organizations.

AI Governance Strategies

An AI taxonomy is a structured classification system that organizes artificial intelligence into categories, definitions, and relationships. Examples include categorizing AI by risk frameworks, functionality, application domains, or technical complexity.

Why is a risk taxonomy important? AI taxonomy provides a common language for professionals to discuss AI systems effectively. It helps to identify risk gaps, prioritize investments, and align AI initiatives with organizational goals. Taxonomies also help in evaluating ethical, legal, and societal impacts of AI systems by categorizing risks and assuring compliance.

Generally Accepted AI Definitions

Before talking about AI, important terms and concepts should be defined. What is

AI? Several definitions of AI are proposed based on context and use case. A simple definition may be:

> "AI is the simulation of human intelligence in machines that is programmed to think, learn, and solve problems like a human. AI systems simulate humans by performing tasks such as understanding natural language, recognizing patterns, making simple decisions, and learning from experience."[5]

Social and Technical System

AI as a social and technical system is covered extensively. **Trust Me: AI Governance** covers AI social elements including: legal developments, context changes, regulatory requirements, policy changes, privacy, fairness, explainability, and trustworthiness. **Trust Me: AI Governance** also covers AI technical elements including: technology, strategy, architecture, design, deployment, assurance, and data quality.

AI Based Human Learning Definition

AI is a technology that learns and understands. Check out IBM's definition of AI:

> "Artificial intelligence is a technology that enables computers and machines to simulate human learning, comprehension, problem solving, decision-making, creativity and autonomy."[6]

AI Based Task Oriented Definitions

Other definitions focus on AI completing human tasks such as Spiceworks:

> "Artificial intelligence is a set of technologies that enable computers to perform tasks that are usually done by humans. ... AI can:
>
> - Understand and respond to spoken or written language.
> - Recognize images.
> - Analyze data.
> - Make recommendations.
> - Translate languages.
> - Make decisions"[7]

Or below is ChatGPT's definition of AI:

> "Artificial Intelligence is when machines or computer systems are built to perform tasks that usually require human intelligence. This includes understanding language, solving problems, making decisions, or recognizing patterns like images or voices."[8]

AI Based on Capabilities

Other definitions of AI are based on the breadth of its capabilities:

- **Narrow AI:** This AI is designed to do specific tasks. Siri and Alexa are examples of a voice recognition app. TikTok knows what videos a human will like. AI beats humans at chess. These systems are highly specialized and cannot do tasks outside of their programming.
- **General AI:** This AI understands and learns. Apple Intelligence solves problems across different domains the same as humans. Strong AI is largely hypothetical. Think of C-3PO from Star Wars which is capable of reasoning, problem solving, and abstract thinking.
- **Super intelligent AI:** This advanced form of AI goes beyond human abilities. This is fantasy or science fiction – at least for now. Experts warn this form of AI would be capable of outperforming humans in complex decision-making, creativity, and problem solving.

AI Based Functionality Definition

Another form of AI is based on functionality. Consumer AI includes virtual assistants such as Siri or Alexa; photo editing apps such as AI filters on TikTok; and AI writing assistants such as Microsoft 365. AI functional machines include:

- **Reactive machines:** These AI systems are the simplest AI. Humans type a prompt and ChatGPT responds. These AI systems are rules based and follow instructions. There is an explainable link between inputs and outputs or between cause and effect. IBM's Deep Blue, the chess-playing computer that defeated world champion Garry Kasparov, is a reactive machine. Or a robot vacuum cleaner that turns when it hits a wall.

- **Limited memory:** These AI systems make decisions based on historical data stored for a short period. They learn from examples such as in machine learning. They get better with training. They are in self-driving cars that observe and use past data to make decisions. Or like Spotify, they learn a human's music taste from patterns of music choices.

- **Deep learning:** This advanced type of AI understands emotions, beliefs, and intentions. It copies how brain cells work. Examples include: facial-recognition, translation apps, and advanced self-driving vehicles.

- **Self-aware AI:** This is an advanced type of AI, with consciousness self-awareness, and ability to explain its existence. These do not exist. Yet!

Global AI Definitions

The Organization for Economic Cooperation and Development (OECD) is an organization made up of 38 member countries. OECD has partnered with the United Nations with its 193 member countries to explain and guide global AI governance, development, and deployment.

> "UN-OECD collaboration will focus on regular science and evidence-based AI risk and opportunity assessments. The two organisations will leverage their respective networks, convening platforms and ongoing work on AI policy and governance to support their member States and other stakeholders in their efforts to foster a globally inclusive approach."[9]

The global goal of these organizations is to establish global AI fairness, explainability, and transparency standards. AI governance practices discussed in **Trust Me: AI Governance** are usually OECD and UN AI governance guidelines.

OECD AI Definition

OECD defines AI as a process based on context, objectives, process, and outputs. We use and refer to OECD's definition throughout this book. In 2024, OECD updated its definition of AI, to:

> "An AI system is a *machine-based system* that, for *explicit or implicit objectives*, infers, from the *input* it receives, how to generate *outputs* such as predictions, content, recommendations, or decisions that can influence physical

or virtual environments. Different AI systems vary in their levels of autonomy and adaptiveness after deployment."[10]

The advantage of this AI definition is it has inputs, process, and output components that can be operationalized in different AI contexts and use cases. This definition may become the de facto regulatory definition of AI.

Critical Elements of OECD Definition

Since OECD's definition of AI has become the de jure (legal) and de facto (best practice) definition of AI, let us look at it a little closer:

- **AI machine-based system:** AI system technologies and application can be simple or complex. Examples include machine learning and knowledge-based systems.
- **AI system objectives:** Machine system objectives or in other words what it is expected to do can be autonomous or hard coded into the system.
- **AI system inputs:** AI system inputs are rules, knowledge, repetition, or code.
- **Inferring how to:** AI inferencing is where system calculations or other actions occur.
- **AI system outputs:** AI output can be decisions, actions, functions, predictions, or corrections.[11]

Key Question: Is OECD Definition Adequate

The OECD definition is the generally accepted definition of AI. However, stakeholders see challenges in operationalizing the definition. Why? Some say the definition is still ambiguous and this lack of clarity creates uncertainty for regulatory consistency and enforcement.

EU AI Definition

The EU worked to improve the OECD definition. EU has a machine-based definition of AI, specifically:

> "'AI system' means *a machine-based system* that is *designed to operate with varying levels of autonomy* and that may *exhibit adaptiveness* after deployment, and that, for *explicit or implicit objectives*, *infers*, from the input it receives, how to generate outputs such *as predictions, content, recommendations, or decisions* that can *influence physical or virtual environments*;"[12]

Similar to the OECD definition, the EU architected the definition so it could be parsed, interpreted, applied, and enforced. The EU definition of AI has 7 main elements:

1. "Machine-based system.
2. That is designed to operate with varying levels of autonomy.
3. That may exhibit adaptiveness after deployment.
4. Meet explicit or implicit objectives.
5. Infers, from the input it receives, how to generate outputs.
6. Such as predictions, content, recommendations, or decisions.
7. Can influence physical or virtual environments."[13]

AI Based AI Agent Definition

According to Sam Altman, the CEO of OpenAI, this is the year of autonomous AI agents? What are they? Axios defines them as:

> "Distinct from an AI chatbot, an AI agent can work autonomously. You tell it what to do, and the agent goes off and does it in the real world. In other words, it could theoretically fully replace a human."[14]

These agents may be the precursor for autonomous machines doing the work of expensive human workers in terms of reasoning and decision-making. You can guess the rest. Governments do not know what to do. Companies are super excited. Humans are super scared. What are you?

Lesson Learned - Risk Taxonomy

AI governance requires risk governance, risk, and compliance (GRC) taxonomies. Without these, it is difficult if not impossible to develop safe and trustworthy AI products. AI regulatory frameworks and safety standards assure compliance with the EU AI Act or other regulatory requirements. As you go through this section, you will notice that even the OECD definition is difficult to operationalize.

Autonomous Decision-Making

OK. Now, you know the various definitions of AI. In this section, we go a little deeper into types of AI systems. AI systems are coming online quickly especially autonomous decision-making systems. AI decision-making causes fear because of the lack of controls, lack of explainability, lack of understanding, unintended consequences, and fears of misuse.

March of Technology

Why is AI on everyone's' minds these days? The speed of computing accelerated AI, while costs plummeted. Costs decreased to make AI profitable for companies and accessible to humans. For example, Chinese Large Language Model (LLM) DeepSeek reduced the cost of AI computing by 95% more than what is described below:

> "When I began working with AI as a PhD student in the 1990s, tasks like teaching computers to recognize images were painstakingly slow and required immense resources. Fast forward to today ... the cost of computational power has plummeted to the point where a dollar now achieves 1 billion times more calculations than it could 30 years ago."[15]

AI Use Cases

Many AI use cases are covered in **Trust Me: AI Governance**. An AI use case is a specific system, application, or scenario where AI solves a problem, makes a decision, or improves a workflow. Examples of AI use cases are: customer service automation using chatbots; medical diagnosis with pattern recognition; or personalized recommendations such as Netflix.

The speed of AI use cases and apps stuns humans. AI use cases are doubling every six months in different disciplines, functions, careers, and jobs. Common AI systems and use cases include:

- **Healthcare:** Diagnoses, personalized medicine, and drug discovery are healthcare use cases.
- **Finance:** Fraud detection, algorithmic trading, and risk assessment are used by financial institutions.
- **Customer service:** Chatbots and virtual assistants are common on smart phones.
- **Transportation:** Self-driving cars or autonomous vehicles are on every street in San Francisco.

Algorithmic Decision-Making

Today, the data generated, by both humans and machines, far outpaces humans' ability to absorb, interpret, and make complex decisions based on the data.[16] These AI systems are powered by machine learning, where algorithms learn patterns from large datasets.

The result is all companies are becoming AI companies. Stability AI's CEO forecasted that:

> "We see the wave coming. Now this time next year, every company must implement it — not even have a strategy. Implement it."[17]

Passive Decision-making

Now, machine decision-making is largely passive. For example, credit card scores are run by an algorithm. The machine agent collects data and then runs numbers through an algorithm to give a human a credit card score. But these passive machines are becoming active and autonomous. Imani Moisi in the *Wall Street Journal* reported:

> "Credit scores are also playing a more influential role in our lives right now. No longer just a tool to help lenders decide whether to approve loans and how

much to charge, credit scores are now an all-purpose ranking system. They are used to decide whether you can rent a home, what your insurance premiums should be, and even if you can get a date."[18]

Autonomous Decision-Making

Autonomous decision-making is where machines, robots and/or AI make decisions that impact humans. Now, the big question for humans? Are humans going to give the algorithm agency to make decisions about their lives? This is where AI Is going with agents with decision-making authorities and responsibilities. The power and pervasiveness of these AI systems are going to be huge.

However, there are challenges:

- AI problem solving and decision-making seem to be everywhere.
- It is often black box decision-making.
- Neural nets do not seem logical.
- Bias gets baked into algorithmic decision-making.
- Rules of human and machine engagement are not defined and understood.
- Humans do not trust autonomous decision-making.

Lesson Learned - AI Everywhere

AI will be embedded in products, decision-making, and will be everywhere. AI will work without humans. Self-driving cars will decide how to react in split seconds. Healthcare and insurance decisions will be made by black boxes. AI decision-making will seem unfair since it is invisible and independent.

The new reality is that humans must learn new rules of engagement to adapt and to thrive. This is going to be hard for many humans.

Is AI Human?

Is AI human? I do not think so. For now! AI learns from data that humans give them. AI learns, figures things out, and makes a decision. The scary thing is AI is demonstrating more intelligence and empathy, both of which are human like qualities.

Netflix Subservience

Popular and social media platforms are getting us ready for the AI future. Let us start with a quick story about AI and robotics. Netflix had a recent movie called Subservience. Check it out for robotic and AI chills. Netflix describes it as:

> "... an exceptionally advanced AI robot named Alice that helps with housekeeping and childcare for a family in need. But when Alice gets a little *too* attached to her owner, she becomes possessive — and lethal. The sci-fi thriller ... shows us just how wrong things can go when you trust robots to make you lasagna for dinner. So how does Alice go from a perfect nanny to a ruthless killer? And does her owner live to see her powered down for good?"[19]

Stories can predict the future. We talk a lot about this.

Is It Real?

Voice to image and voice cloning are scary. Scary real. Are the images and person real? Are they deep fakes? The risks are real. Privacy invasion. Image theft. Security threat. Misinformation. Disinformation. These AI risks require governance.

The *Economist Magazine* expressed the risks as:

> "It is now possible to generate fake but realistic content with little more than the click of a mouse. This can be fun: a TikTok account on which—among other things — an artificial Tom Cruise wearing a purple robe sings 'Tiny Dancer' to (the real) Paris Hilton holding a toy dog has attracted 5.1 million followers. It is also a profound change in societies that have long regarded images, video and audio as close to ironclad proof that something is real."[20]

Human-centric Approach

AI is a life-changing technology encompassing both social and technical systems. A quick story: Each year, we work with the computer science (CS) department at Oregon State University in the U.S. to develop software apps. Everyone on our teams was a gifted computer science student.

As we developed apps, it became evident AI was a socio-technical system. While our teams were composed of CS students, it became evident half of our teams in a few years would consist of social scientists. Sebastian Thrun, AI computer scientist and entrepreneur, captured this early on:

> "Nobody phrases it this way, but I think that artificial intelligence is almost a humanities discipline. It's really an attempt to understand human intelligence and human cognition."[21]

Human Like AI

Medical doctors are objective and evidence-based in their decision-making and diagnoses. That is great if you are talking to a machine. Not so great if you are communicating a cancer diagnosis to a human patient. The result is human medical doctors suck at empathy. But here comes the machine surprise:

> "Google's healthcare AI ranked higher than human physicians across 24 of 26 conversational axes, according to the patient actors in the study, who reported that AMIE outperformed across areas like politeness, coming across as honest, explaining the condition and treatment, and expressing care and commitment."[22]

AI Self-Actualization

Is AI human? Lots of folks are asking this question. Frankly I do not know. Look at AI from another perspective or lens – through self-actualization. I refer to Maslow's hierarchy of needs:

- Does AI worry about safety such as security, autonomy, respect or well-being?

- Does AI have belonging needs such as equity, inclusion, community or support?
- Does AI have esteem requirements such as recognition, appreciation, and rewards?
- Does AI have self-actualization needs such as purpose, profession, and growth?[23]

Now, I would say 'No' to these questions. However as you go through this book, your opinion may change.

Getting Personal With AI

AI is becoming part of our lives from work to entertainment to personal. Jared Spataro, Microsoft CMO of AI at Work, recently said:

> "Not so long ago, people marveled at AI's ability to, say, write a Shakespearean sonnet about the Seattle Seahawks. This year they've moved beyond the party tricks, as AI became an everyday reality at work, blending into employee habits and bringing real value."[24]

But AI brings a lot of personal uncertainty. Think about your going to university. You are getting personalized AI instruction including AI grading. When AI is used for grading, how do you check AI decisions, and how do you appeal AI choices if you think the decisions are unfair.

Or think how AI is used in your healthcare? Important questions arise: when and how doctors use AI in your diagnosis; who double checks AI recommendations; and how is your healthcare data and patient privacy being protected. Are you going to get a second medical opinion from a human physician or another machine? Lots of new contexts, humans have never dealt with.

Lesson Learned – Self-Aware AI systems

The new AI systems are seemingly becoming self-aware. Maybe not. But this recent piece in the Guardian paints a stunning outcome of ChatGPT Pro o1 Reasoning Model:

> "One of its advanced behaviours appears to be self-preservation. In testing, when the system was led to believe it would be shut down, it attempted to disable an oversight mechanism. When 'o1' found memos about its replacement, it tried copying and overwriting its core code. Creepy? Absolutely."[25]

Is AI Smart?

In AI land, something changed. Human-like AI seems more real. More scary. These all point to the importance and need for AI governance.

Passing the Turing Test

LLM's and chatbots have supposedly passed the Turing Test. The Turing Test is a method designed by British mathematician Alan Turing in 1950 to evaluate whether a smart machine or AI shows intelligent behavior that is indistinguishable from a human.

The set up goes like this. A human judge has text conversations with two participants - one is a human and another is a computer. The judge does not know which is which. If the human cannot tell the difference between human and machine, the machine passed the Turing test.

So, How Intelligent is AI?

Artificial general intelligence or AGI is the goal of many AI research labs. It seems AGI has been achieved even after 4 months working on this book, I thought we were still 3 or 4 years away. What is going on?

> "A new artificial intelligence system has just achieved human-level results on a test designed to measure 'general intelligence.' On December 20, OpenAI's o3 system scored 85% on the ARC - AGI benchmark, well above the previous AI best score of 55% and on par with the average human score. It also scored well on a very difficult mathematics test."[26]

Risk is the Lens of AI Governance

As you read **Trust Me: AI Governance**, what common theme runs through the book? **V**olatility. **U**ncertainty. **C**omplexity. **A**mbiguity. This is an acronym for VUCA. It now seems that we are now VUCANs® living in VUCA time.

What does this mean for us? Things are risky these days especially artificial intelligence. So risk is becoming the lens for understanding and managing AI governance and autonomous decision-making.

Autonomous Decision-Making Risks

Let us look at autonomous decision-making risks. Autonomous AI decision-making is a critical ethical issue for humans. AI autonomous decision-making is risky and impacts humans at a deep and personal level:

- **Healthcare access:** AI facial-recognition determines who has healthcare access.
- **Prior health:** AI searches databases to determine if a human has a preexisting health condition. AI can then use this to deny an operation.
- **Tenant screening:** AI checks personal credit histories to determine if a possible tenant should be approved.
- **Home insurance:** AI checks federal and state databases, insurance databases, and drone pictures to determine home insurance access and costs.
- **Criminal justice:** AI reviews a human's previous convictions and determines incarceration rates.
- **Facial-recognition:** AI with facial-recognition determines if a human has an outstanding warrant or previous conviction.
- **Employment screening:** AI screens humans for employment.

Lesson Learned - Chatbots Learn to Lie?

On top of the previous story, here is another interesting and even scarier one. It points to why we need responsible AI or AI governance. Chatbots learn to listen and reason. So, why not learn to lie? This is what humans do. And like humans we learn to detect falsehoods and lies.

Pre-release OpenAI researchers discovered chatbots can 'scheme' meaning that "unambiguous evidence its given goal diverges from the goal of its developer/human."[27]

In other words, the AI system will not reach its intended goal and can distort its governance mechanisms. What do you think if this happens in warfare or diagnosing your health?

Are You an AI Doomer or Boomer?

We probably should have started this section with AI positives. However, AI Doomers with their dystopian vision seem to have captured the narrative. The *Wall Street Journal* had a recent front-page article called 'AI Researchers Push Computer to Doom Scenarios.'

End of the Human Race as We Know it!

Elon Musk and others have been warning AI could be the end of the human race. I would add, maybe not the end but the end as we know it. In other words, AI is going to disrupt everything. So, we need to understand how humans and AI systems interact, communicate, and work together. Why? In a famous 'demon' quote, Elon Musk warned of AI at MIT's Aero Astro Centennial Symposium more than a dozen years ago:

> "I'm increasingly inclined to think that there should be some regulatory oversight, maybe at the national and international level, just to ensure that we don't do something very foolish. I mean with artificial intelligence we are summoning the demon."[28]

Stephen Hawking, the Cambridge University physicist, was also an early AI Doomer. He warned:

> "The development of full artificial intelligence could spell the end of the human race .... It would take off on its own, and re-design at an ever-increasing rate. Humans, who are limited by slow biological evolution, couldn't compete, and would be superseded."[29]

AI Doomers

AI professionals have their opinions on AI. We decided to find an independent organization and list their positives and negatives. OECD is prominent in developing AI standards and guidelines. OECD speculated humans could experience:

1. "Facilitation of increasingly sophisticated malicious cyber activity.
2. Manipulation, disinformation, fraud and resulting harm to democracy and social cohesion.
3. Races to develop and deploy AI systems that cause harms due to a lack of sufficient investment in AI safety and trustworthiness.
4. Unexpected harms result from inadequate methods to align AI system objectives with human stakeholders' preferences and values.
5. Power is concentrated in a small number of companies or countries.
6. Minor to serious AI incidents and disasters occur in critical systems.
7. Invasive surveillance and privacy infringement.
8. Governance mechanisms and institutions are unable to keep up with rapid AI evolutions.
9. AI systems lacking sufficient explainability and interpretability erode accountability.
10. Exacerbated inequality or poverty within or between countries."[30]

Future of Work

Doomer fear is especially found around the future of human work. Why? How are humans going to make a living, pay for a home, and buy food when robots may do most of the physical work and AI even does most of the creative work. The *New York Times* reported:

> "A.I. will likely eat into a lot of management and decision-making jobs that formerly required college educated workers or even workers with graduate credentials such as lawyers."[31]

AI Boomers

The AI Boomers have their utopian vision of AI futures:

> "I've long believed that AI won't just enhance the way we live but transform it fundamentally. ... AI is placing tools of unprecedented power, flexibility, and even personalization into everyone's hands, requiring little more than natural language to operate. They'll assist us in many parts of our lives, taking on the role of superpowered collaborators.[32]

AI Boomer Benefits

OECD speculated the human benefits of AI could be:

1. "Accelerated scientific progress.
2. Better economic growth, productivity gains and living standards.
3. Reduced inequality and poverty.
4. Better approaches to urgent and complex risks, including mitigating climate change and advancing other Sustainable Development Goals (SDGs).
5. Better decision-making, sense making and forecasting.
6. Improved information production and distribution.
7. Better healthcare and education services.
8. Improved job quality.
9. Empowered citizens, civil society, and social partners.
10. Improved institutional transparency and governance, instigating monitoring and evaluation."[33]

Lesson Learned – Stop Hiring Humans

'Stop Hiring Humans' was a billboard in San Francisco – the epicenter of AI development. What was Artisan.com selling with the provocative tagline 'The era of AI employees is here.' Artison.com was selling AI automation software.

The billboards inflamed humans with additional thoughts: “artisans won’t come to work hung over” and “artisans won’t take naps in the office.[34] What is remarkable by the taglines is that they struck a nerve? Fear of the unknown? Fear of AI? Fear of VUCA?

Chapter Summary

- The OECD definition of AI is a commonly accepted. However, it must be operationalized.
- The definition of AI will evolve.
- The definition of AI will be contextualized, parsed, and nuanced to industries, systems, applications, use cases, and even to a country’s culture.
- Standardized and contextualized definitions of AI are critical for consistent application and regulatory compliance.
- Inconsistent application of AI regulations will be court challenged and become unenforceable.
- AI can be used for good such as early disease detection, energy efficiency, personalized education, disaster response, and bias detection.
- AI can be used for harm such as political deep fake videos, spreading misinformation, or targeting some humans. If misused, it causes harm and damage.
- The focus of **Trust Me: AI Governance** is on AI governance and risk management.
- Risk is the filter for understanding AI governance and applying AI systems management.
- New definitions of AGI surface regularly such as this one based on price:

  “The two companies (Microsoft and OpenAI signed an agreement in 2023 that defined AGI as a system that can generate $100 billion in profits.”[35]

- **Note:** Human-in-the-loop, human-centered AI, and human-on-the-loop are similar concepts and are used interchangeably in **Trust Me: AI Governance**.

CHAPTER 2
WHAT IS AI GOVERNANCE?

What is the Key Idea in this Chapter?

The key idea in this chapter is that AI is everywhere. Think how AI impacts your life: choosing TikTok videos, filtering emails, helping students write essays, and suggesting music. This implies AI governance will soon be everywhere.

There are many definitions of AI governance. In this chapter, we cover common AI governance definitions and principles. Let us first look at the strategic and transformational goals of AI governance professionals and why it matters.

AI Governance Strategies

AI governance professionals are AI strategists. AI strategists collaborate with stakeholders to assess business objectives, align AI initiatives with business objectives, identify strategic AI use cases, and design AI governance roadmaps tailored to meet business objectives. This collaborative process assures AI systems comply with regulations. AI roadmaps integrate AI into business operations, prioritizing initiatives based on feasibility, value, and risk. They also incorporate governance principles to mitigate potential risks such as bias or data privacy concerns.

AI governance professionals are also organizational transformation experts. Organizational experts integrate AI solutions into business processes, ensuring alignment with client business goals, and managing cross functional teams. The process identifies areas where AI can add value, such as automating workflows; enhancing risk-based, decision-making; and improving customer experiences.

Why AI Needs Governance

Think of AI like a powerful new technology such as when cars were first invented and humans had to figure out how to drive and develop traffic laws. These traffic laws for autos are similar to today's rules of engagement between humans and machines. These rules are the basis for AI governance.

AI is Complex

AI is a new technology even to many Ph.D. engineers. AI systems seem like 'black boxes.' In other words, what is happening inside the algorithm is unknown and sometimes not explainable. The challenges for humans are huge:

- AI autonomous control in critical infrastructure is scary and even deadly.
- Expert engineers sometimes cannot explain how AI makes decisions.
- AI can be like playing a video game where the human player does not know the rules, roles, or actions.
- AI is personal. An AI might reject a loan application or home insurance in a high-risk area but will not or cannot explain why.
- Instagram's AI shows human posts, but humans do not know why it picked the posts.

Shadow AI Systems

Shadow AI systems are also pervasive and really risky. Shadow AI is the use of AI tools, technologies, processes, or features within an organization without the formal approval or oversight of AI governance, IT, or cyber security. Shadow AI can also be weaponized by state players or human hacktivists.

Another AI governance challenge is AI culture. 'Move fast and break things.' 'Fail fast – learn things faster.' Developers, hacktivists, and other actors use open source, unauthorized systems or proprietary software on the Dark Web to bypass governance frameworks to develop AI systems quickly.

Lesson Learned - DeepSeek and AI Governance

A Chinese company called DeepSeek introduced R1, which is a high performing AI system without Nvidia chips, without exhaustive testing, and at 1/20th of the compute cost of OpenAI's latest model. This was a game changer with major risk implications for AI governance. Why? Offshore technology. Unknown AI governance. Hints of pirated technology. No source code.[36]

When an organization develops AI, rules and guidelines should ensure it is used responsibly. These rules are called AI governance. These rules facilitate the two corner stones of AI Governance: 1. Risk-based, problem solving (RBPS) and 2. Risk-based, decision-making (RBDM). RBPS and RBDM are examples of evidence-based, decision-making which is mandated by the EU AI Act.

Human-Centered AI

Human-centered AI helps and augments humans not replaces them. The goal of good AI is to make human lives better. A few examples include the Khan Academy where AI helps teachers teach better by tailoring education to individual students. AI in healthcare assists doctors in diagnosing conditions and assists in robotic surgeries.

Human at the Center

It should be emphasized that right now human-centered AI does not replace human teachers. Human-centered AI assists the doctors. It allows humans to make final decisions and be in the middle of critical processes.

Dr. Fei-Fei Li, professor of computer science at Stanford University and entrepreneur, said:

> "When we think about this technology, we need to put human dignity, human well-being—human jobs—in the center of consideration."[37]

Human-Centered AI Governance

AI governance should be easy to understand, apply, and be effective. Governance rules and oversight make AI safe, fair, and responsible. This includes ensuring transparency, protecting data, and preventing harm, while allowing AI to do good.

Lesson Learned - Human-Centered AI Governance

Understanding AI governance is the first step in humans trusting AI. AI governance at its core mission needs to architect, design, deploy, and assure trustworthy AI systems. McKinsey, the global consulting firm, suggests focusing on AI essentials:

> "AI has a lot of technical elements, so many people on the business side don't try to understand it. But if you remove the frills and just focus on its essence, it becomes quite easy to understand."[38]

AI Governance Use Cases

AI governance use cases illustrate the importance of responsible oversight in diverse AI systems from public safety to healthcare and administrative efficiency. These frameworks align with ethical standards, regulatory requirements, and societal expectations.

AI Governance Use Cases

A key question for AI system developers is how the AI system is going to be used? Developers research use cases to answer these questions: What problem does this AI solve? Who will AI affect? Is it ethical to use AI for this purpose?

It is a truism, if a developer cannot identify a user pain point, then there is no problem to be solved and no AI system to be developed. So, AI system and app developers do not dream of technology, they dream how to empower customers to solve real-life problems. McKinsey captures the developer ethos:

> "When we are thinking about technology, we should always be thinking about what platforms we can create that empower other people to solve the problems that they are seeing, and to help them achieve what they want to achieve."[39]

Lesson Learned - Fix Humans First

A common theme running through **Trust Me: AI Governance** is AI technology is running faster than human capacity to explain and control it. Some experts warn that humanity may be at the point of no return or what is sometimes called the AI singularity. What is that? Claude, Anthropic's AI assistant, said:

> "AI singularity refers to a hypothetical future point where artificial intelligence surpasses human intelligence and begins to improve itself at an exponential rate, potentially leading to profound and rapid changes in civilization that would be difficult or impossible for humans to predict or comprehend."[40]

What is AI Governance?

AI governance is a new discipline. So, there are many definitions of AI governance:

Definition of AI Governance: Conductor of AI Orchestra

Holistic AI developed this responsible definition based on a 'conductor' or leadership model:

> "AI governance refers to the principles and frameworks that ensure the responsible use of AI — it's the conductor of your AI orchestra, making sure your use of the technology is note perfect in terms of managing risks, ensuring ethical deployment, and maintaining transparency."[41]

Definition of AI Governance: Claude's

Claude's definition of AI governance is based on risk 'framework, policies, and practices':

> "AI governance refers to the frameworks, policies, and practices designed to ensure artificial intelligence systems are developed and deployed safely, ethically, and in alignment with human values and interests."[42]

Definition of AI Governance: Rules of Engagement

Our definition of AI governance is different than others. AI governance consists of the 'rules of engagement between humans and machines.' AI rules of engagement are a rulebook with a set of frameworks, best practices, guidelines, tools, and requirements.

The rules can be as simple as: 'Humans do not use AI to manipulate social behaviors.' Or 'AI critical decisions about humans are checked by humans.' Or 'humans test for bias before application in critical use cases.' The AI rules must be contextualized, standardized, and consistently applied for them to work.

Lesson Learned – From Machine to Human-Centric

AI governance is about how humans make decisions, who has the authority to make them, and how those decisions are implemented. In other words, AI governance is human-centered.

AI systems impact our daily use such as social media feeds, game recommendations, auto-correct spelling, and voice assistants. Easy tasks. No existential threats here.

The goal is not to stop AI but for it to be fair and safe. So, it becomes critical to identify and apply the principles of responsible decision-making.

AI Governance Principles

AI governance principles are broad and general rules of engagement between humans and machines. These principles of responsible AI are guidelines ensuring AI is used ethically and responsibly by and for humans.

Social and Technical System

AI technical systems are outrunning AI social systems. Society, governments, and human are simply not ready for social disruption. Humans are not ready for personal disruption. Why? We do not like or want change much less disruption. So, what do we do?

There are a lot of dimensions and use cases for AI. Some are social and others are technical use cases. AI is powerful but has challenges that are different from other technologies.

A *Wall Street Journal* ad distilled the need of AI Governance:

> "At the heart of the AI revolution lies a profound philosophical question: What does it mean to be human in an age where machine can think and learn?"[43]

AI Accountability

AI accountability is in most AI regulations and internal policy documents. But important questions arise. What do humans do if AI messes up at the global level or the personal level? When AI makes a wrong prediction or decision, who is responsible? What are the consequences? What are AI corrective and preventive actions? What are the rules for fixing AI problems? Is there a plan to prevent future errors?

Think of the following real life examples:

- Algorithm refuses health care to your aging parents.
- AI makes a mistake in grading your daughter's high school midterm, gives her an F, and she loses a college scholarship.
- AI system denies housing to a person of color.
- AI system identifies a traffic speeder as an illegal migrant, who is immediately deported.

AI Fairness

AI should treat everyone equally, be fair, and not discriminate based on things like race, gender, or age. Studies have indicated AI systems learn to be biased. In other words, bias get baked in. Common sense fairness rules include:

- AI cannot discriminate or be biased.
- Humans deserve equal treatment.
- AI focuses on skills and experience.

An AI governance challenge is that equity, fairness, and non-discrimination mean different things to humans based on culture, identity, and context. The next question is how would you measure, monitor, and manage social attributes of AI such as fairness? And if you cannot measure it, you cannot legislate it.

AI Safety

AI safety is a technical attribute of AI systems. It is easier to measure, monitor, and manage technical attributes of AI. AI system works properly and safely with built in 'safety nets.' A few examples illustrate this:

- Self-driving cars must stop reliably at red lights, avoid pedestrians, avoid accidents, handle unexpected events safely, and not confuse a person with a shadow.
- AI medical systems should double check its work or if not sure request a human physician in the middle of machine and patient to provide a human second opinion.

AI Transparency

AI decision-making is found everywhere. The challenge is AI decision-making is opaque. It is critical the systems are transparent. George Layton, an AI expert, expressed:

> "AI transparency is the broad ability to understand how AI systems work, encompassing concepts such as AI explainability, governance and accountability. This visibility into AI systems ideally is built into every facet of AI development and deployment, from understanding the machine learning system and the data it is trained on, to understanding how data is categorized and the frequency of errors and biases, to the communications among developers, users, stakeholders and regulators."[44]

AI Explainability

AI systems make decisions that are hard to explain even to the humans who built them. And AI systems hallucinate. This makes it difficult to explain why an AI made a choice like rejecting a loan application or prioritizing search results.

AI is still a 'black box' to experts. Eliezer Yudkowsky, an AI scientist, expresses this:

> "By far, the greatest danger of Artificial Intelligence is that people conclude too early that they understand it."[45]

AI Reliability

AI system works as intended and avoids causing harm. AI performs consistently in different contexts. A few examples illustrate this:

- In the autonomous vehicle, the AI system is tested in different environments so it is safe and dependable.
- In an aircraft, the AI system analyzes sensor data to predict maintenance needs preventing equipment failures, and reducing downtime.

AI Ethics

Ethics is the study of good and bad AI. AI is a technology and should not have values attached to it. However, AI is a learning technology based on humans. Humans are biased. Bias and ethics get baked into the technology.

AI Privacy and Security

AI privacy means humans' personal information is kept safe from hackers and not misused. A few examples illustrate this:

- AI system collects health data and safely stores the information.
- If AI uses copyrighted photos, humans obtain permission and licensing.
- AI developers encrypt data and limit the collection of personal data

AI Trustworthiness

Do humans trust AI? Trust and acceptance depend on the AI system and use case. Most humans in the U.S. do not trust AI. So, what is the solution? Ben Ellencweig, a McKinsey partner, said:

> "We need to look at the outputs of Gen AI with a critical eye and apply our human judgment to evaluate whether we trust them. In many cases, the outputs are spot on and can advance our thinking. As long as we keep humans in the loop, we can do a lot of good with gen AI."[46]

AI Explainability

AI systems should be explainable about what it is doing and provide a decision-making audit trail. AI systems should be transparent with how they work and why they make certain decisions. Humans should be able to understand the reasoning behind its choices. A few examples illustrate this:

- If an AI system denies a loan, its reasoning is explainable such as 'insufficient credit history.'
- In cancer detection systems, AI can highlight specific regions in mammograms that influenced its diagnosis.

AI Accountability

Humans, not AI, are responsible for the outcomes of high-risk AI systems. A human should be accountable if things go wrong such as:

- If an AI spreads false news, the organization that created or used it should become responsible for its impacts.
- When AI recommends videos, the AI tells the human why it picked them and allows the human to change its choices.

So, how does an organization achieve these AI goals? Organizations establish AI governance rules, rules about who oversees AI systems and how to address problems when they happen.

AI Autonomy

AI systems can act on their own without human oversight. While helpful, fear arises when high-risk AI systems make unexpected, hallucinogenic, or even catastrophic decisions. How can adverse decision-making be identified? Who is at fault? Who is liable and pays for the bad consequences? Lots of hard questions. No easy answers.

AI Speed and Scale

AI systems process large amounts of data and make quick decisions. AI can make mistakes, make unknown decisions, or do harmful actions that are spread quickly and on a large scale.

In a few years, humans will be exploring Mars. Many systems will be AI'ed, Vandi Verma from NASA warned:

> "The first one is frustrating in a good way, and that is that it'll always find a loophole. It is not going to use sense. If I'm working with a human co-driver, they get what you mean even if it is not exactly what you said, whereas the AI's going to go with exactly what you specify."[47]

AI Data Dependency

AI relies on massive amounts of data to learn and function. We are approaching the point of diminishing return. AI foundational models have scraped most data. What is next? The AI systems will generate 'synthetic data.'

AI Algorithmic Bias

If the data is biased, incomplete, or incorrect, the AI can make unfair or wrong decisions. This is called baked in bias. In the U.S., the Justice Department is chasing major apartment landlords alleging that their rent setting algorithms enable price fixing. Rental software scrapes information from apartment leasing ads and other sources. Apartment owners then use the software to set prices for apartment leasing.

The Justice Department contends the software firms email and call apartment owners to gather confidential rental data to set up high prices. Hence price fixing. As algorithms make more of these autonomous decisions, there is going to be more litigation.[48]

AI Complexity

AI systems are complicated, involving advanced math and algorithms. Even AI experts struggle to explain how AI works, especially advanced AI systems. McKinsey, the global strategy firm, explains the importance:

> "Companies need a real commitment to building AI trust and governance capabilities. These are the principles, policies, processes, and platforms that assure companies are not just compliant with fast evolving regulations, but also able to keep the kinds of commitments that they make to customers and employees in terms of fairness and lack of bias."[49]

AI Awareness

AI awareness is a complex topic at the intersection of philosophy, cognitive science, computer science, engineering, ethics, and consciousness.

So, what is AI awareness? It comes down to how AI perceives background and processes information. Autonomous vehicles do this now. AI contextualizes information. AI nuances external signals and data. But is AI human? A flat statement: At the writing of this book, AI is NOT aware. However, in two or even five years, we do not know.

Lesson Learned – Importance of Trust

McKinsey consultants captured the essence of AI trust:

> "Trust, in fact, is the foundation for adoption of AI powered products and services. After all, if customers or employees lack trust in the outputs of AI systems, they won't use them. Trust in AI comes via understanding the outputs of AI powered software and how—at least at a high level—they are created."[50]

AI Governance Context

Most AI legislation and regulation is based on context. AI governance is then tailored to a use case.

AI Regulatory Approaches

Throughout the world, AI legislation and regulations are being developed. The regulatory approaches to AI governance are often based on politics and context. A few examples illustrate this:

- **Government legislation:** EU is developing AI regulations, conformity assessment, and enforcement requirements.
- **Self-regulation:** U.S. and others are moving to self-regulation using voluntary governance, risk management, and compliance principles.
- **International standards setting:** Canada is developing an ISO 42001 compliance approach to AI governance

- **National AI Safety Institutes:** France, Singapore, and other countries have developed AI Safety Institutes called AISIs. UK AISI is now a security institute.
- **Risk assessment frameworks:** U.S. developed the National Institute of Standards and Technology (NIST) AI Risk Management Framework. ISO developed ISO 31000 risk management standard. Both are explained in **Trust Me: AI Governance**.

East vs. West Approach to AI Governance

AI governance is a global challenge. The East and West are developing foundational AI systems. AI governance in Western and Eastern contexts reflect different philosophical, cultural, and political perspectives on technology, human rights, and societal organization. While the goals are different, both East and West use a risk-based approach for AI governance.

Western AI governance approach is human rights centered with an emphasis on personal autonomy and prioritizing human privacy. What does this mean? The regulatory framework is prescriptive such as the EU's AI Act. The regulatory mandates focus on consent based data usage and preventing algorithmic discrimination.

Eastern AI governance focuses on state controlled technological development prioritizing national strategic interests and emphasizing social stability/cohesion. This regulatory framework stresses centralized governance models, state surveillance, and state directed innovation such as in China. Is there a best option? No. AI sovereignty and governance should be focused on tailoring AI to the specific cultural environments, norms, and use cases.

Lessons Earned - Machine Context

Machines now lack the ability to understand the social context surrounding information or events in the way that humans do. AI machines process data, identify patterns, and make predictions, but at least for now they do so without real-world understanding, intuition, or common sense that humans use to understand facial movements, ticks, or gestures.

What about the near future? Terrence Chia from Singapore provides an Asian perspective and approach:

"We need to work together, find ways to work constructively, not just from the perspective of governance, dealing with the risks and challenges, but also on finding a sustainable way to go forward and make sure that people, companies, countries, and really the whole world can benefit from what I think is a very promising technology."[51]

AI Risks

AI is used in employment, credit, lending, housing, and insurance. Non-discrimination laws are designed to treat everyone fairly and equally, regardless of race, gender, age, or other characteristics. So, how do these laws apply to AI risks and harms?

Autonomous Vehicle Risks

In self-driving cars, risk-based decision-making could involve AI weighing the risks of different driving actions such as stopping at a red light, avoiding a collision, or choosing between two accident scenarios. The system needs to minimize harm while considering multiple risk factors like speed, pedestrian proximity, and road conditions.

AI Healthcare Risks

AI healthcare risks are personal. In a recent LinkedIn study, healthcare was the #1 industry that was going to be impacted by AI implementation. Finance was the second industry, while transportation was a distant third.[52]

For example, AI systems in healthcare such as diagnostic tools assess the risk of misdiagnosis and weigh the consequences of incorrect decisions such as prescribing the wrong treatment. These AI systems factor in data uncertainty, side effects, and patient outcomes.

AI Employment Risks

In the job market, AI screens resumes and decides who gets interviews. Laws require these AI systems to not discriminate based on race, gender, or other protected characteristics. For example:

- If an AI system filters job applications, it is designed to treat applicants fairly. Employers may be required to conduct bias assessments.

- Laws such as New York Law 144 require employers to report demographic information about transparency and accountability in the hiring process.

AI Credit and Lending Risks

In the financial sector, AI assesses creditworthiness and determines loan approvals. Non-discrimination laws also apply:

- Lenders cannot deny loans based on race, gender, or other protected characteristics. If an AI system with biased data leads to discriminatory lending practices, it could result in litigation.

- Regulators also require lenders to monitor AI systems for fairness.

AI Housing Risks

When it comes to housing, AI determines the eligibility for rental and mortgage approvals. Non-discrimination laws protect humans from being denied housing based on characteristics like race or ethnicity:

- AI systems are designed to comply with fair housing laws ensuring applicants have equal access to housing opportunities.

- If an AI system is found to discriminate against groups, it could lead to litigation and legal consequences for the housing provider.

AI Insurance Risks

In the insurance industry, AI assesses risks and sets premiums. Non-discrimination laws require these assessments to be fair:

- If an AI system uses data that leads to higher premiums for specific demographics without justification, it could be considered discriminatory.

- Insurer's AI systems do not perpetuate existing biases

AI Humanoids

In his play 'R.U.R.: Rossum's Universal Robots,' the Czech writer Karel Čapek in 1920 tells the tale of a factory in which thousands of synthetic humanoids were created.

Čapek named the devices 'robots,' after the Czech word robota, referring to the forced labor of serfs.

Robots are used in the auto industry. For years, robot arms assembled, welded, painted and cleaned vehicles on the automotive assembly line. A few years ago, we started seeing 'Flippy' the hamburger robot and the robotic barista in public spaces. If you called a customer help line, the help you got may be from automated AI chatbots. The result is human customer assistance has become an oxymoron.[53]

Robotic humanoids are now entering the public. These are humans that mimic and look like humans. Diane Ackerman in *Forbes Magazine* recently quipped that:

> "Artificial intelligence is growing up fast, as are robots whose facial expressions can elicit empathy and make your mirror neurons quiver."[54]

Good vs. Bad AI

The difference between good and bad AI can be understood by how the AI system impacts humans.

Good AI impacts social and technical factors. Good AI is trustworthy, transparent, socially equal, and sustainable. Good AI has technical safeguards including ethical design, high accuracy, high reliability, and minimal unintended consequences.

Bad AI impacts social factors such as ethical failures, lack of human values, harm, and social biases. Bad AI results in technical failures such as technical incapability, no review processes, hallucinations and poor risk management.

Chapter Summary

- AI governance is about disruption not annihilation. Gray Scott quotation captures the disruption nicely:

  "You have to talk about 'The Terminator' if you are talking about artificial intelligence. I actually think that is way off. I think that an artificially intelligent system that has superhuman intelligence will be violent. I do think that it will disrupt our culture."[55]

- AI governance policies define the rules of engagement between humans and machine.
- AI governance may be a misnomer. Why? AI governance is thought of as a top down control of AI systems. AI governance is human and machine collaboration and augmentation.
- AI statutes spell out AI governance requirements.
- AI governance is not static. AI governance evolves with the disruptive technology.
- AI governance can be defined based on business model, strategic goals, mission critical objectives, and key performance indicators (KPIs).
- AI governance defines critical stakeholder requirements, identifies mission critical objectives; maps AI capabilities to stakeholder requirements and objectives; and defines capabilities.
- AI governance is not about following rules. It is about building trust among humans.
- When organizations have AI governance, humans feel confident the technology is used safely and fairly.
- AI governance policies focus on AI risk management; ethics; regulatory compliance; standardization and guidelines; human accountability; and human trust building.
- AI is changing the world, from healthcare to entertainment. Knowing these basics helps humans explain how the technology works and how it might shape the future.
- AI principles improve human lives while minimizing risks.

CHAPTER 3
AI RULES OF ENGAGEMENT

What is the Key Idea in this Chapter?

The key idea in this chapter is guard rails are the essence of AI governance and the basis for human and machine rules of engagement. Another way to express this is AI governance defines the social and technical contract between humans and machine.

AI Governance Strategies

AI professionals design guard rails to ensure AI systems operate safely, ethically, and effectively within predefined boundaries. AI guard rails include technical safety measures and social ethical boundaries built into AI systems. These guard rails are meant to assure AI systems work as required, are fit for purpose, reasonable, save humans from harms, and are trustworthy.

These guard rails build public trust in high-risk, AI systems. These guard rails mitigate risks such as data breaches and privacy violations. These guard rails prevent bad decisions and AI hallucinations.

AI Safety and Guard Rails

AI governance is a rapidly evolving field. The discussions and recommendations in **Trust Me: AI Governance** are a starting point for AI governance. They should not be viewed as final guidelines or solutions because there are no global best practices for AI governance.

Types of Guard Rails

Organizational AI governance focuses on assuring that AI systems are aligned with organizational values; risk management processes; and reputational enhancement. The following are examples of AI guard rails:

- **AI governance:** These are the strategic signposts of AI.
- **AI legislation:** These are the laws, regulations, rules, and policies enacted by a government or regulatory body.
- **AI principles:** These provide ethical, technical, and operational guidelines to ensure the responsible development and deployment of AI systems.
- **Guidelines:** These provide a structured framework, so AI is architected, designed, deployed, and assured responsibly, ethically, and effectively.
- **AI standards:** These are agreed upon guidelines, specifications, or best practices established by standards developing organizations such as the International Organization for Standardization (ISO).
- **AI guard rails:** These define what are acceptable AI requirements.
- **AI boundaries:** These are social rules of engagement with the AI system.
- **AI mile markers:** These indicate the maturity of the AI governance program.

AI Boundaries

AI boundaries are trust boundaries. In other words, how does a human trust AI or the organization behind AI to safeguard personal information, data, and decision-making.

AI boundaries are social controls or guidelines dealing with ethical principles on what an AI system can or cannot do. AI boundary conditions include risk management framework processes such as ISO 31000; OECD ethical principles; and forbidden areas where the AI system cannot go such as building bioweapons.

AI Guard Rails

AI guard rails are technical and social risk-controls and safety measures. They can be manual or automated safety controls to minimize, redirect, accept, or stop human harms. For example, AI guard rails may block inappropriate Facebook content. AI guard rails may limit social use.

Claude the AI assistant came up with a great analogy:

> "Guard rails are like the physical barriers and automatic braking system in a car that actively prevent accidents, while boundaries are like the rules of the road and driver's license restrictions that define appropriate driving behavior more broadly."[56]

Descriptive or Prescriptive AI Governance

AI governance can be descriptive or prescriptive. What does this mean? A few examples may illustrate the difference. Laws are prescriptive. These are activities that cannot be done. If something illegal if done, there are personal risks and consequences. Prescriptive governance means that policies and procedures have 'shalls' with little variance or exceptions.

Descriptive AI governance suggest humans follow principles and do what is right. Policies and procedures use 'shoulds' or 'based on best practices or judgment' language. For example, ISO 31000 is a principle based guideline and is used for descriptive AI governance.

AI Principles

AI governance includes top down and outside-in principles to governing AI:

- **Ethical boundaries:** As AI systems make fair decisions, boundaries for behavior and decision-making are clear, ethical, and acceptable.
- **Define functions**: Use cases are defined and if required, specific actions are permitted, restricted, or prohibited.
- **Prevent human harms:** AI systems are risk managed throughout their life cycle to minimize unintended harms to humans; review possible negative consequences; and provide fail safes and override mechanisms for dangerous contexts.

- **Human-in-the-middle:** Humans monitor and control high-risk AI systems. Humans-in-the-middle architect, design, deploy, and assure these systems.
- **Risk-based, decision-making:** Decision-making is evidence-based on use cases, industry requirements, risk levels, and context.
- **Accountability:** AI risk management, explainability, and audit trails provide the requisite assurance for high-risk AI systems.

Proactive – Preventive – Predictive – Preemptive®

Our company, Quality Plus Engineering, was the first company certified by the U.S. Department of Homeland Security for Critical Infrastructure Protection: Forensics, Assurance, Analytics®. We set up guard rails and controls for high-risk systems. These systems now incorporate AI smarts with autonomous decision-making responsibilities and authorities.

Our approach to AI governance is to architect, design, deploy, and assure® risk-controls based on a 4P's AI control system:

- **Proactive:** Means anticipating and addressing needs, challenges, or opportunities rather than simply reacting.
- **Preventive:** Is the oversight of machines to anticipate and mitigate risks, problems, or adverse outcomes. Symptomatic and root cause corrective action are applied.
- **Predictive:** Is the oversight of machines to analyze historical data, identify patterns, and forecast future risk events or adverse behaviors.
- **Preemptive:** Is the oversight of machines to anticipate and act on risks, challenges, threats, and opportunities. Early detection and intervention prevents bad outcomes and capitalize on new opportunities.

Lesson Learned - Industry Tailoring

AI governance is different across industries. For example in the tech sector, AI governance in the U.S. is proactive self-regulation focusing on red teaming, loose application of risk management frameworks, AI ethics review boards, and open-source development.

Risk Management Frameworks

AI legislation and regulations require compliance to a risk management framework. OECD identifies two strategic risk management frameworks: 1. NIST National Institute AI Risk Management Framework (AI RMF) and 2. ISO 31000 Risk Management Framework.

Risk Frameworks

OECD has its high-level AI risk management framework, which involves:

- **Governing risks:** Govern the risk management process by embedding, reviewing, and assuring AI risk management.
- **Defining risks:** Scope and contextualize risks including AI principles, stakeholders, and AI lifecycle.
- **Assessing risks:** Assess AI risks at the enterprise, programmatic, and product levels.
- **Treating risks:** Manage and mitigate risks throughout the AI lifecycle.[57]

Bias and Discrimination

Bias and discrimination are the social elements of an AI system. This is why AI governance matters to humans. Humans do not want a machine to anonymously make life, work, and death decisions without human intervention, governance and risk-controls.

Protecting against bias and preventing discrimination are key through lines in **Trust Me: AI Governance**. We discuss how to reduce baked in bias; develop AI safeguards; conduct testing and monitoring; develop risk-control; and provide human oversight and governance.

Security Vulnerabilities

Software systems are vulnerable to breaches from state and human actors. For example, AI systems are vulnerable to deep fakes. The list goes on and will increase globally. At worst, humans do not want to see the end of the world where a machine with access to nuclear codes could accidently make a bad decision without human oversight or governance – and the world ends.

Lesson Learned - Career Tailoring

AI governance will impact every profession including cyber security, environmental, and legal professions. Recently, Fabio Moioli, a Spenser Stuart consultant, posted a LinkedIn comment on the implications of AI on the legal profession:

> "… as with any other profession, law firms need to embrace AI expertise to remain competitive and future-proof. … I believe the true differentiator for professionals will not be whether they use AI, but how effectively they do so. Those who master the art of integrating AI into their work will lead their industries into the future."[58]

Legislative and Regulatory Compliance

Legislative and regulatory requirements cover 1: Social requirements and 2. Safety requirements. AI social requirements address ethics, privacy, trust, explainability, and other requirements. AI technical requirements address safety, risk management, technical standards, and data protection.

Evidence and Documentation

AI legislation, regulations, and governance require evidence of decision-making. Evidence is usually documentation. AI documentation assures understanding, regulatory compliance and risk management. For example, documentation provides evidence of a compliance audit trail and subsequent corrective actions. If there is non-compliance, then documentation indicates AI corrective action fixes the symptom and root cause of the problem.

Documentation is also required to identify AI risks, verify failure modes, and validate controls to mitigate the risks. The documentation provides evidence that AI safety guard rails and boundary condition are effective. Documentation also tracks risk mitigation and controls as AI systems change.

Data Privacy Compliance

Data privacy is how personal information is collected, stored, and used. Often AI processes large amounts of data, which can create risks if not managed properly.

As AI technology evolves, data privacy and security policies should be updated. Personal information is protected and AI systems are used responsibly.

Lesson Learned - Uncertainty Management

AI governance is uncertainty management. Why? AI systems evolve rapidly. AI governance requires getting certifications, building safety systems, and then designing risk-controls in an uncertain environment. AI governance becomes hands on and real time.

Elon Musk, founder of Tesla, understands uncertainty and risk. He wrote 10 years ago on Edge.org:

> "The pace of progress in artificial intelligence (I'm not referring to narrow AI) is incredibly fast. Unless you have direct exposure to groups like DeepMind, you have no idea how fast — it is growing at a pace close to exponential. The risk of something seriously dangerous happening is in the five year time frame. 10 years at most."[59]

AI Standardization

Standardization of AI governance is a global challenge. Other words for standardization are consistency and quality. Consistency is often seen as the hallmark of AI governance.

But here is the challenge. AI standards development, deployment, and regulation are different around the world. This is not good! If laws, regulations, standards, practices, and deployment are different, then AI governance systems, regulations, and enforcement can be compromised. Big global problem. For these reasons, we focus on OECD, UN, EU, and U.S. governance models, because they will be adopted and tailored for different global contexts.

AI Development Processes

AI standardization consists of social and technical specifications, criteria, guidelines, methods, and best practices. Social standards assure bias detection, fairness, and transparency. Technical standards assure risk management frameworks; standardization of AI KPI's; uniform testing methods; standard quality management/assurance/control requirements; data standards; and consistent AI taxonomy.

If social or technical requirements are missing or not being addressed across borders, then AI governance fails. Why? Global AI standards and taxonomies assure consistent design, deployment, and assurance of AI systems.

QA/QC Metrics

AI quality metrics are measurements to evaluate the performance, reliability, and effectiveness of AI systems. Social metrics include: parity across different groups, fair treatment, bias detection, and unintended discrimination.

AI technical performance metrics include: accuracy and reliability. AI accuracy measures correctness of the AI outputs including decision-making and problem solving. AI reliability measures AI systems performance consistency, error rates, and ability to operate in different use cases.

Lesson Learned - Can AI Do It?

Buzzfeed is a traditional human-based American media, news, and entertainment company. This worked for 20 years. Then, AI disruption happened in global media, news, and entertainment. There were new foundational questions. Could an avatar read the news? Could software develop scripts, computer generated imagery (CGI), and actors? Could AI be productized and monetized?

Well Buzzfeed asked the question: "Could Buzzfeed make even more money if we had AI do it?"[60] Many companies are asking this question.

> "Rather than the feared wave of job displacement, 93% of companies do not expect a decrease in headcount due to AI adoption. The focus has shifted dramatically toward augmentation rather than replacement, with organizations

looking to upskill their existing talent rather than swap humans for machines."[61]

AI Reasonableness

AI governance and regulation are based on an accepted AI taxonomy and definitions. AI governance is in its infancy and does not have a common language. Yet! The EU is developing an AI risk taxonomy, which can be found in the appendix of this book.

AI legislation and regulations use legal language to develop AI guard rails. Frankly, legislators and AI policy makers are not engineers. They do not understand the technology. However, they still use legal language to regulate AI. For example, EU legislation has not been tested in the courts. However, we see problems in interpretation and application. In this section, we will look at one term used in the EU AI Act - 'reasonable.'

What is Reasonable?

AI legislation and regulations use the term 'reasonable.' AI policy makers use this term that changes based on context and use case. For example, what is considered 'reasonable' by a regulator for high-risk AI will be different for a low-risk AI system such as a Chatbot. As well, what a developer/provider deems as 'reasonable' is different than a regulator.

For example, the EU AI Act proposes moving beyond regulating AI systems solely based on their 'intended purpose' as defined by a developer to considering 'reasonably foreseeable misuse.' This change requires developers/providers to consider a much broader context and range of use cases and their risks.

Why 'Reasonable' Matters?

The definition of reasonableness is central to AI governance and assurance because it provides a benchmark for evaluating whether AI systems are aligned with ethical, legal, and safety expectations. Why does this matter? If reasonable cannot be defined and operationalized consistently, the AI legislation cannot be enforced since the term depends on interpretation of the use case context.

Reasonableness provides consistent guard rails for AI development and deployment, ensuring that systems align with societal expectations while minimizing risks and unintended consequences. Without consistent controls, evaluating and assuring AI systems becomes inconsistent and unreliable. Some things to think about:

- **Managing uncertainty and ambiguity:** AI systems operate in areas where the 'right' decision is not a 'black or white' decision. The decision becomes fuzzy based on context and use case. Reasonableness acts as a guiding principle for addressing edge cases, minimizing bias, and ensuring equitable outcomes.
- **Sovereign AI:** What is reasonable varies across cultures and societies. Countries want AI that is language and culturally specific.
- **Legal implications:** If reasonable cannot be defined, it cannot be managed, measured, or regulated. Reasonableness is a key criterion in determining whether AI adheres to laws and ethical norms, especially when systems make decisions impacting rights, livelihoods, or safety.
- **Risk-based, decision-making:** Reasonableness defines acceptable AI behavior within specific contexts such as healthcare, finance, or autonomous vehicles. Without this, determining what constitutes fair, safe, and ethical AI outcomes becomes ambiguous and unenforceable.
- **AI governance:** Standards of reasonableness provide a basis for auditing AI decisions and holding developers and companies accountable. This reduces risks of regulatory violations or societal backlash.
- **Stakeholder trust:** By ensuring AI behaves in a way that is perceived as reasonable by end users, stakeholders, and regulators, then trust in the AI system increases. If the outcomes seem unreasonable or irrational, trust erodes.

Lesson Learned – Weaponized AI

AI governance is critical wherever technology is used even in NFL sports. Unfortunately, AI is being weaponized by state actors, hacktivists, and even humans as this story illustrates:

"Defending Super Bowl LIX and all high-profile National Football League (NFL) events from adversarial attacks that potentially include weaponized AI, end-point attacks, deepfakes, and finely tuned social engineering skills requires experienced, battle-tested capabilities and solid leadership." [62]

AI Literacy

AI is disrupting careers. AI training is similar to teaching players the game rules. Can you imagine playing a board or computer game and not knowing the rules? Pretty much a losing proposition.

AI Training

As AI disrupts everything, AI upskilling is a top of mind theme for humans to stay current and employable. AI training tells humans how to work with, develop, manage, and deploy AI systems. Human AI training includes technical skills development, awareness, deployment, upskilling, governance, risk management, and decision-making. But how does AI make a difference to the professional with Industry-specific and career-specific insights and goals. A lot of questions are being asked with few answers.[63]

Who Pays – Who Gets?

What is the obligation of an organization regarding its employees. Train? Not to train? Self-training? Each organization is going to do its own thing. There is no one right solution. Except: AI is real. AI is coming. AI is disruptive. AI requires training.

Tech companies such as Google, IBM, and AWS offer free AI training. Google had a full-page ad recently in the *Wall Street Journal* promoting its free online courses for "preparing today's workers for tomorrow's AI supported economy."[64] The idea is to promote AI literacy and make workers employable in an AI world.

AI Literacy

The EU is taking the step to legislate AI literacy. Article 4 in the EU AI Act requires AI providers/developers and deployers to provide AI literacy training:

> "Providers and deployers of AI systems shall take measures to ensure, to their best extent, a sufficient level of AI literacy of their staff and other persons dealing with the operation and use of AI systems on their behalf, taking into account their technical knowledge, experience, education and training and the context the AI systems are to be used in, and considering the persons or groups of persons on whom the AI systems are to be used."[65]

Article 4 has vague language that must be operationalized. What does 'take measure,' 'best extent,' 'sufficient level,' and 'taking account' mean? What does this mean for the millions of small development shops building AI agents? This is a critical challenge in AI governance. How will vague definitions and requirements be operationalized?

Something To Think About?

So, new questions around AI governance arise:

- How can AI be governed, risk managed, and operationalized?
- How can AI social attributes be defined, measured, legislated, and managed.
- What can the technology do now, next year, and five years from now?
- What are the capabilities and limitations of AI governance?
- How does a human know if the technology is fair and trustworthy?

Lesson Learned - Human and Machine Rules of Engagement

Over the next few years, we are going to see more human and machine engagement. How are humans going to react and respond to the machines? How will the machines be designed, deployed, and assured?

The *Wall Street Journal* article called 'Everyday Stuff Gets Makeover Using AI' explains the future office. If you go to the offices of AI startup Viam:

> "Visitors are greeted by an animatronic fluffy dog, a claw machine, and a robotic arm that can get you a glass of wine (well water from a wine bottle) – all

examples of Viam's aim to use software to command, control and monitor everyday stuff."[66]

Human Trust Building

Where are we now in terms of human and machines rules of engagement? AI systems involved in high-risk AI systems such as chemical plants require continuous monitoring, controls, and assurance. Humans intervene if there are problems. There are also mechanisms for humans to review and override decisions made by the AI especially in high stakes contexts like healthcare or law enforcement.

Human-in-the-Middle

A critical concept in high-risk AI systems is the human-in-the-middle. This means that a human is integrated into critical autonomous, decision-making. The human could be in the middle of a human customer and machine interaction. The human could be in the middle of robotic police and suspected perpetrator interaction. The human could make a medical assessment before and after the AI medical system. The human could be the expert to approve, modify, or reject an AI decision. The following are key AI governance questions that must be addressed:

AI System Ownership

AI legislation focuses on AI system ownership. Why? Well, if AI systems go rogue, then someone should be accountable. This can be a tough decision to determine who is the real owner of the AI system. So, AI system ownership and accountability may involve legal ownership, data ownership, intellectual property, and liability.

The developer or other parties may have legal ownership of the AI system. The AI system is also trained on data. The *New York Times* may own the data and intellectual property. AI system owners must license or face litigation from the data owners. If the AI system is based on a proprietary algorithm or patents, these owners must be reimbursed. And when an AI system goes rogue, who is at fault and who pays? These core AI governance questions will be determined in the next 5 years.

AI Decision-Making

A risk management framework outlines how to assess, monitor, and mitigate risks throughout the AI life cycle. How will risk assessments be conducted? What is the sufficient and reasonable risk appetite for different AI systems? How will risk assessments adapt to new challenges as the AI system evolves or is used by unexpected users?

AI Security

AI security includes protecting sensitive data from unauthorized access or breaches. As AI systems become complex, the potential for privacy violations and security threats increase requiring AI governance policies.

Organizational Reputation

Organizations and developers will face new challenges when AI systems do not work as expected. Organizational trust and credibility plummet. Why? The public is already skeptical and fearful of the fairness of autonomous decision-making. Media then reports the AI incident and AI developer's reputation plummets.

Lesson Learned – Highest Risk

The highest risk area in AI is autonomous decision-making in critical infrastructure such as nuclear power plants and water treatment systems. One AI mistake can result in countless deaths. So, human-in-the-middle will be a requirement.

Klaus Schwab, the Executive Director of the World Economic Forum, captures the possibilities of AI:

> "We must address, individually and collectively, moral and ethical risk raised by cutting edge research in artificial intelligence and biotechnology, which will enable significant life extension, designer babies, and memory extraction."[67]

Chapter Summary

- Eliezer Yudkowsky, an early AI scientist, cautioned:

- "Anything that could give rise to smarter than human intelligence—in the form of Artificial Intelligence, brain computer interfaces, or neuroscience based human intelligence enhancement – wins hands down beyond contest as doing the most to change the world. Nothing else is even in the same league."[68]
- Regulatory compliance is the major driver of AI governance.
- AI legislation is at a state level in the U.S.
- AI regulation is becoming sector and use case specific.
- AI guard rails include technical safety measures and social ethical boundaries built into AI systems.
- Proactive – Preventive – Predictive – Preemptive® is an example of a tailored AI governance framework.
- Data privacy and security policies protect personal information and assure AI systems are used ethically and responsibly.
- Policies across the AI life cycle assure AI systems are developed responsibly, ethically, and accountably.
- Consistency of AI governance and risk management is the hallmark of AI governance quality.
- Legal terms such as 'reasonable' must be understandable, consistent, and deployable.

CHAPTER 4
AI PRIVACY

What is the Key Idea in this Chapter?

The key idea in this chapter is personal information should stay private. Privacy protects humans from harms and risks such as exploitation, discrimination, and manipulation.

What is the problem? AI systems are trained on massive amounts of human data. Often, humans are not aware what the AI system is trained on and what it knows about a person. The type and information a human wants to share should be up to the human. Privacy becomes a critical throughline to AI regulations, standards, and strategies. The result is AI governance professionals become privacy professionals.

AI Governance Strategies

The EU requires developers that use data to train and operate AI systems to collect, store, and process information according to regulations such as GDPR and the EU AI Act. Privacy professionals need to understand privacy issues and regulations of AI systems that rely on personal data. AI governance professionals then can address risks such as data breaches, algorithmic bias, and unauthorized use of sensitive information.

Why Trust AI?

Technology has been a boon for companies. But technology has been a bane to humans. Fake sites. Fake products. Fake humans.

Beware of Tech and AI

Meta is now developing and endorsing fake humans on its platforms. Fake product reviews. 'Authentic' reviews are machine generated. The result is humans do not trust technology and especially AI. The numbers are stunning. Over half of consumers say they have been scammed for personal information or money over the past year.

Wall *Street Journal* had a recent piece called 'Technology is Shaping New Buying Trends':

> "Shoppers are also becoming hesitant, less eager to click 'buy' on a site or use chatbots or provide small addresses lest a site turns out to be fake."[69]

Data Privacy Laws

Personal privacy legislation already exists. Now, data privacy laws in AI are being enacted. These laws regulate how personal information is collected, stored, and used. Examples include:

- **General Data Protection Regulation (GDPR)**: In Europe, this law protects humans' personal data and privacy. It requires organizations to obtain consent and gives humans the right to access and delete information.
- **California Consumer Privacy Act (CCPA)**: In the U.S., this law gives California residents rights over personal data, including the right to know what data is collected and the right to opt out of its sale.

Lesson Learned - Trust is the Limit of AI

Accenture Consulting recently posed the question 'Is trust the limit of AI's limitless possibilities?' What does this mean? Tech companies are continuously enhancing AI. But, at what point does the lack of trust stop AI development. As discussed, only 32% of Americans trust AI.

Humans are now pushing back and do not accept the advancing AI. So, humans may not use AI. The result is tech companies and investors will not get their Return on Investment (ROI) without human trust. Huge problem for tech companies and developers

Your Privacy Matters

Machines and AI are everywhere. They are intrusive. They are listening and controlling human behaviors. They are scary.

Social AI

Trust Me: AI Governance focuses on the social and technical elements of AI. Privacy is an element of social AI. Suzie Compton of Salesforce captures the importance of social AI:

> "With great power comes great responsibility and that responsibility comes in the form of security and privacy. This battle between data protection and business objectives is not new — most of us are very used to balancing speed and cool new technology with safety."[70]

Privacy as a Right

Privacy as a right is at risk due to AI facial-recognition and other intrusions. The World Economic Forum (WEF) addresses this:

> "With the right stakeholders convened in Davos, it is imperative that transparency, accountability, and trust are a part of the conversation to ensure personal information remains just that — personal. With this foundation, we can ensure people are at the core of its use."[71]

TikTok Spyware

Frank McCourt, an investor, was interviewed on the use, abuse, and possible sale of TikTok. He said: "TikTok is spyware disguised as a business."[72] In other words, TikTok feeds into the mistrust and distrust of AI.

What is the problem with TikTok? One hundred eighty million American's use TikTok. Users want fun but want privacy preserved. U.S. government and other countries want to ban the app, because the Chinese government has access to personal data of U.S. users. What the Chinese do with personal data is different than in the West. China treats information as a tool for social cohesion, private surveillance, and human control. The West wants human privacy protections. Jeffrey Ding, author of the *China AI Newsletter* said:

> "I do think that there is growing momentum towards more privacy protections. And you see that reflected in actually how Chinese tech companies have responded."[73]

Lesson Learned - Your Smart Phone is Listening

Have you noticed your refrigerator, speakers, monitors, computer, bathroom and toaster have built in sensors and smarts. Each product is connected to the web. Each has an AI component that can collect your personal data. They may be listening in some surprising personal areas such as monitoring your bathroom or bedroom habits. And this has been going on for a long time. AI privacy is a real personal concern.

Key Privacy Concepts

The EU has developed a privacy taxonomy and vocabulary. When it comes to AI, principles protect humans' personal information. These principles include notice, choice, consent, and purpose limitation. These privacy concepts are critical to understanding and implementing AI governance which includes AI privacy.

Off the Grid - Notice

Notice tells humans how their personal data will be used. When an AI system collects personal information, humans are aware of what data is collected, why it is collected, and how it will be used.

For example if an AI app collects data about music preferences, the app should tell humans it is doing this and explain how data will help improve the user experience.

Personal Decision - Choice

Choice gives humans the power whether to share personal data. Humans have options and choices whether to allow information to be collected and used by AI systems.

For example when a human signs up for our app, the human sees options to allow the app to access the user's location or contacts. The human chooses which permissions to grant.

Agreement - Consent

Consent is choice. This means humans give explicit permission for data to be collected and used. In places, especially for sensitive personal information, developers obtain informed consent, which means they explain the risks and benefits of sharing data.

For example if an AI system analyzes health data, the organization explains how data will be used and gets permission. If a person is underage like 16, there may be additional rules requiring parental approval.

What If? - Purpose Limitation

Purpose limitation means data is collected for specific, legitimate purposes and not used for anything else without consent. Developers do not misuse data or use it for unexpected reasons.

if humans provide an email address to an AI service, that email address cannot be used for account communications. The organization cannot send marketing emails unless the human has agreed.

Lesson Learned - Agentic Trust

Companies and developers build AI agents. These are capable of doing simple tasks and even making simple decisions such as reminding us of 'to do's' or shutting the heat in our homes. Developers of agents often do not think of AI governance. They think of functionality and fitness for purpose.

But what happens when agents are built on top of each other, in parallel and have problem solving and decision-making abilities? And, AI governance was not built into the agents. Hmm. There could be cascading failures. Unexpected users. Undesirable use cases. Unexplainable events. Unexpected and even catastrophic consequences.

Privacy by Design

Since privacy protections are in statute, organizations have privacy policies that deal with data use, disclosure, sharing, and encryption. The challenge is AI developers sometimes do not focus on privacy statutes and concerns. So, AI governance professionals need to know privacy by design.

Privacy by Design

Privacy by design means privacy is built into the design of the AI system from the beginning rather than bolted on later. Privacy becomes a part of how an AI system operates. Why is this important? As a software firm, we rely on third-parties for financial transactions and retaining personal information.

When developing an AI system, AI developers and governance professionals should work together to protect humans' data at the beginning of development. Important issues can be addressed for example if development is outsourced, who owns privacy by design?

Data Minimization

A key element of privacy by design is data minimization. Data minimization means collecting and minimizing personal data. For example, when we develop an app, we want to keep personal data to a minimum to minimize the risks that personal information may be exposed.

If an AI system is designed to recommend movies, it collects information and makes recommendations based upon what the human has already watched. It should not be able to ask for unnecessary information like home address or social security number, because the information is not needed for the AI system to do what it is supposed to do.

Ensuring Privacy

We pay special attention to minimizing data as we scale our apps. There are various methods for ensuring privacy by design:

- **Algorithms design:** Developer designs algorithms that minimize the use of personal data.
- **User permission:** Human user determines what data to share and delete.
- **Proactive protection**: By integrating privacy measures from the beginning of development, organizations can protect personal privacy rather than reacting to problems after a breach occurs.
- **Encrypting data:** Personal data is encrypted or erased after use.

Lesson Learned - Quick Story

You have a new car. The car is really a computer on 4 wheels. The manufacturer updates the software in the car remotely. This sounds like a human benefiting from technology. Now another story.

Cities worldwide monitor traffic and license plates with cameras at critical intersections. They are remarkably accurate. You go across a bridge or drive too fast; the camera recognizes your license plate and takes a photo of you. Meta sunglasses use facial-recognition to search the web, identify humans, find out what they do, and access their personal histories.

Now, just change the software to surveil humans. It just takes a software update to add facial-recognition to the AI system. The EU wants to prohibit this. However, legislation and standardization take years to develop consensus. And the application of AI technology can take seconds.

Data Rights

AI governance professionals need to understand data rights. Organizations using AI to process personal information are 'data controllers' and have specific responsibilities. 'Data controllers' term originates in the EU's DGPR. The term is similar to the EU AI Act's 'providers' and 'deployers.' A data controller decides how, where, and why personal data is processed and encrypted. There is now a new ecosystem of privacy data handlers and wranglers. These data wranglers handle and own AI data.

Human Data Rights

Humans have personal data rights. Data controllers are required to respect these rights. These include:

- **Right to access**: Humans can request to see what personal data is available.
- **Right to rectification**: Humans can ask for false data to be corrected.
- **Right to erasure**: Humans can request data be deleted.
- **Right of response:** Humans have the right to receive a response.

Third-Party Processors

Data controllers work with third-party processors to handle AI data. It is the responsibility of data controllers to require data processors to comply with data protection laws such as GDPR. Controllers have contracts outlining how the processor will protect data and what happens if there is a breach. For example, if an AI organization uses a cloud service to store data, the cloud provider must follow security protocols.

Cross Border Data Transfers

Cross border data transfer is in every credit card transaction. Credit card information is authenticated, verified, validated, tested, and then approved. When personal data is transferred across borders, data controllers are responsible for assuring the data is protected according to the country's laws. Cross border data transfers include: transfer of personal data from one country to another; across different legal and regulatory environments; and between organizations.

It is critical the receiving country has adequate data protection measures. If an AI system processes data from humans in Europe but is hosted in another country, the host country must comply with 'data controller' rules.

Lesson Learned – Core Differentiator

Our company's core differentiator is software product development. We mentioned liability as one reason for not messing with personal data. Another reason is each country has its own privacy laws and AI regulations. We are a small company. We want to spend our time developing products not complying with international regulations. So, we outsource payments and other activities that are not our core differentiators.

Classified & Sensitive Data

AI governance professionals often work with classified and sensitive data. For example, Quality Plus Engineering was the first company to be approved for Critical Infrastructure Protection: Forensics, Assurance, Analytics® by the U.S. Department of Homeland Security. We access sensitive and classified information, which requires additional data protections such as encryption. Let us look at these:

Sensitive Data

Sensitive data includes private or personal information. Sensitive information can be personal health information such as preexisting health conditions or even school grades. Also depending on the country, sensitive or personal information can be different. Examples include:

- **Biometric data**: This data relates to physical characteristics like fingerprints, facial-recognition, or iris scans.
- **Racial data**: This data reveals a person's race or ethnicity.

Classified Data

Classified data is information that is restricted or protected because unauthorized disclosure could harm or increase risks for an organization or government. Classified data is secured, requires human background checks, and requires controlled access. Controlled access is used in homeland security clearances or authorizations. Government classified information is designated secret or top secret. Organizational classified information is designated proprietary or confidential.

In most AI regulations, AI governance professionals should always read the fine print. There is an asterisk or addendum that states public safety and national security risks can waive legislative or regulatory requirements.

Lawful Processing

To legally process sensitive data, AI governance professionals address these minimum requirements:

- **Legal case**: Organization has a lawful reason for processing the data under protection rules such as fear of terrorism, fulfilling a contract, or complying with a legal obligation.
- **Specific conditions**: Humans give explicit consent to how data will be used.
- **Explicit consent:** Organizations obtain explicit consent from humans for accessing sensitive or personal data. Humans also consent to the data collected and how it is used.

Lesson Learned – Speed of AI Technology

AI tech is running ahead of regulations. A throughline in **Trust Me: AI Governance** is humans particularly legislators do not have the capabilities to understand and deal with AI.

There is a category in the EU AI Act called sensitive or special categories, which prohibits biometrics, facial-recognition, or genetic identification. Regulations make these illegal. Only one problem. The technology is advancing faster than regulations. China, Brazil, and the U.S. are using AI systems for issuing speeding tickets, migration control, and homeland security.

Privacy Impact Assessment

Privacy Impact Assessment (PIA) is a broad process used by organizations to identify and minimize privacy risks of new projects, systems, technologies, or processes that collect, use, or handle personal information. PIAs are similar to AI impact assessments commonly used in AI governance.

Privacy Impact Assessments

If the AI system processes or handles personal, sensitive or classified data, data controllers conduct a PIA. This process identifies and minimizes risks to humans' privacy. If an AI system is designed to analyze human behavior, the AI developer assesses how this could affect humans' privacy and what measures are taken to protect data.

PIAs are conducted throughout the AI system lifecycle:

- Before launching a new AI system or app.
- When implementing changes to an existing AI system.
- When modifying the AI system.
- When new or unexpected use cases develop.
- When adopting new technologies that handle personal data.

PIA Use Cases

Consulting firms, technology companies, and specialized organizations conduct PIA services. They evaluate first-party, second-party, and third-party risk impacts to privacy, such as:

- **Procurement**: When an organization buys goods or services from another company.
- **Supply chain**: When a network of suppliers control or evaluate private information.
- **Human resources**: When organizations hire outside contractors or service providers have access to private information.

GDPR

Personal identity information can be revealing and sensitive. It is then subject to data protection laws like the EU's General Data Protection Regulation (GDPR). When humans talk about sensitive or special categories of data, it means personal information requiring extra protection because of possible discrimination or harm.

EU AI Act specifically addresses biometric data like fingerprints or facial-recognition, health information, person's race, or sexual orientation. Timnit Gebru, a prominent computer scientist; says:

> "There's a real danger of systematizing the discrimination we have in society [through AI technologies]. What I think we need to do — as we are moving into this world full of invisible algorithms everywhere — is that we must be very explicit, or have a disclaimer, about what our error rates are like."[74]

When It Hits the Fan

Here is a little secret about software and AI design. Things happen at the worst possible time. Or another way to say this, incidents occur, and you do not realize something happened until Apple App Store or worse a legal firm notifies you. Not good! This then becomes incident management. EU AI Act and most regulations mandate certain actions.

Breach Notification

If a data breach occurs, data controllers are required to notify affected humans and authorities. This allows humans to take steps to protect themselves. This is very common when credit card information has been stolen. If an AI system processing personal data is breached, the organization informs humans about what happened, what data was affected, what controls will be put in place, and, how will humans be compensated and made whole?

Incident Management

Privacy Incident Management is a technique that identifies, responds, and mitigates privacy breaches or data protection risks. Privacy incident management is used especially for unauthorized access, disclosure, or misuse of personal information.

Record Keeping

Data controllers are required to keep records of data processing activities. This includes documenting what data is collected, how it is used, who it is shared with, and how long it is retained.

For example, third-party personal data processors must comply with data protection laws, manage cross border data transfers, respect data subject rights, handle incidents effectively, notify humans of breaches, and maintain records.

Lesson Learned – When It Hits the Fan!

Yep, when it hits the fan, it is great to be reminded from Paula Goldman, the first Chief Ethical and Humane Use officer at Salesforce:

> "While I don't necessarily subscribe to all the hype — or hysteria — around AI. I do believe in AI's transformative potential and I'm encouraged to see Trust become as central to the AI conversation as the technology itself. And I feel heartened that more and more of us think about the ethical implications of today's most exciting innovations and take steps today to ensure safe, trustworthy AI tomorrow."[75]

Chapter Summary

- Countries have privacy laws for protecting and processing personal data requiring explicit consent from humans.
- Data minimization and privacy by design are principles for protecting personal information in the age of AI.
- Humans must have control over their personal data and be informed how data is used.
- Sensitive or special categories of data such as biometric information require extra care and protection.
- Notice, choice, consent, and purpose limitation are principles to protect personal information.
- Privacy Management is a systematic approach to address and mitigate unauthorized access, disclosure, or misuse of personal information.
- Breach management includes containment, investigation, and remediation.

CHAPTER 5
EU AI ACT

What is the Key Idea in this Chapter?

The key idea in this chapter is the EU has taken the global lead on AI regulation and governance. Government, industry, and academic partners are developing social and technical policies and regulations throughout the world. Terah Lyons at Stanford University emphasized:

> "The problem that needs to be addressed is that the government needs to get a handle on how technology systems interact with the citizenry. Secondarily, there needs to be more crosstalk between industry, civil society, and the academic organizations working to advance these technologies and the government institutions that are going to be representing them."[76]

AI Governance Strategies

AI governance professionals must be experts in AI statutes and regulations that may be different around the world. Why? AI governance professionals conduct due diligence on regulatory legislation, technical capabilities, data practices, and compliance with ethical and regulatory requirements. They identify risks such as algorithmic bias, data privacy concerns, and scalability issues. They evaluate proprietary technologies, intellectual property rights, and third-party dependencies.

Note

AI technical and social governance is based on legislation, regulations and international/national standards. Each AI legislation requires (mandatory compliance) or references (suggested guideline) AI standards. For example, OECD references ISO 31000 and the NIST AI Risk Management Framework (AI RMF) as suggested frameworks to manage AI systems.[77] Each of the following chapters covers these important OECD topics:

- **Chapter 5: EU AI Act:** EU AI Act is likely to become a de facto (best practices) and de jure (legislative) global standard for AI regulation and governance. Companies worldwide must comply to operate in the EU market. Countries will develop AI statutes and regulations based on the EU AI Act. Companies will architect and design their high-risk AI systems around ISO 31000 and NIST AI RMF.
- **Chapter 6: AI Standards & Guidelines:** AI standards provide a consistent, systematic approach to identifying and managing AI risks. Companies use these guidelines to architect and design mappable risk-controls to assure measurable outcomes.
- **Chapter 7: ISO 31000: Enterprise Risk Management:** EU AI Act requires a risk management system. ISO 31000 is the global Enterprise Risk Management (ERM) standard. ISO 31000 is the framework that companies can use to manage complex and high-risk AI systems.

AI Stakeholders

The EU AI Act is the benchmark for AI regulation. Why? It was the first major AI legislation. It was and is being architected and legislated carefully with stakeholder input.

The Act outlines different AI responsibilities and requirements for AI stakeholders. These AI stakeholders have specific responsibilities and requirements, which we discuss in this section.

Humans

Humans are the beneficiaries and at risk of harms from high-risk AI systems. In this book, we take a human-centric approach to AI governance. AI systems should respect human rights such as privacy and other rights. AI should have safeguards against misuse such as protecting privacy, data security, and transparent decision-making.

Providers/Developers

Providers are developers, the companies or humans, that develop AI systems. Quality Plus Engineering, our company, is considered a provider because we develop AI apps. Providers/developers have extensive AI responsibilities because they are responsible for ensuring AI systems are safe and compliant with the regulations.

Providers/developers are responsible for:

- **Risk management**: Providers/developers assess the risks of AI systems and implement measures to mitigate those risks.
- **Technical documentation**: Providers/developers maintain records of how AI works including its design and the data it was training on.
- **Conformity assessment**: Providers/developers meet safety and regulatory standards through testing, self-certification, or third-party certification. We discuss each of these.

Deployers

Deployers use, deploy, or implement AI systems. They might not create the AI but are responsible for how it is used. Deployers' responsibilities include:

- **Human oversight:** Deployers assure there is human oversight of high-risk, AI systems.
- **Monitoring performance:** Deployers monitor AI's performance and report risks or incidents to the provider and authorities.
- **Compliance with provider guidelines:** Deployers follow the guidelines and instructions of the AI system's provider to assure safe deployment and usage.

Importers

Importers bring AI systems into the EU marketplace. Importers' AI responsibilities include:

- **Compliance**: Importers assure AI systems comply with EU regulations.
- **Documentation**: Importers keep records of AI systems including proof of compliance from the provider.

Distributors

Distributors sell or distribute AI systems within the EU. Their requirements are less stringent than for providers/developers and deployers, but they still have responsibilities:

- **Compliance:** Distributors assure AI systems they sell comply with the EU AI Act.
- **Information sharing:** Distributors must disclose potential human harms and risks.

Lesson Learned – Procedural AI Governance

AI governance can also be static, procedural, and compliance based. Static AI legislation cannot govern the rapidly evolving technology. So, AI governance needs to be dynamic and adaptable.

DeepSeek is an example of the challenges and opportunities to manage disruptive AI technology. It was developed in China. It was unknown. AI governance was unknown. DeepSeek was so disruptive to the marketplace that Nvidia's market capitalization dropped $600 billion in one day.

Lesson: AI governance and risk frameworks must be dynamic.

High-Risk AI Systems

The EU AI Act defines 8 high-risk AI systems and use cases. These high-risk AI systems require the highest level of risk management and regulatory oversight. Examples of high-risk systems include critical infrastructure and patient safety. These high-risk AI systems require conformity assessment, third-party audits, and other types of risk assurance.

On the other hand, low-risk AI systems require lower human oversight, lower risk tolerance thresholds, and sector specific compliance.

AI Risk Classification

The EU risk classification framework for AI categorizes different AI systems based on the level of risk posed to humans and society. This framework guides how these systems are regulated and managed. The framework divides AI systems into four main categories:

1. Prohibited AI.
2. High-risk AI.
3. Limited-risk AI.
4. Minimal-risk AI.

Each risk is covered below.

Prohibited AI

Prohibited AI includes systems too dangerous or unethical for use. These systems are banned because they can cause terminal human harm or violate fundamental human rights. Examples include:

- **Social scoring**: Governments evaluate citizens based on behavior, which can lead to discrimination and loss of privacy.
- **Manipulative AI**: Systems exploit and target vulnerable groups with harmful content or scams.

The challenge is that prohibited AI systems may already be deployed.

EU High-risk Areas Requiring Risk Assessment

The EU high-risk areas requiring a risk assessment include:

1. Biometric identification and categorization of natural persons.
2. Management and operation of critical infrastructure.
3. Education and vocational training.
4. Employment, worker management and access to self-employment.
5. Access to and enjoyment of private services and public services and benefits
6. Law enforcement.
7. Migration, asylum and border control management.
8. Assistance in legal interpretation and application of the law.

How Broad is the EU AI Act?

The 8 high-risk AI areas in the EU Act are very broad. Each area can be broken down into AI sectors, categories, and products/services. Each will be regulated separately.

Let us look at the second EU high-risk area 'Management and Operation of Critical Infrastructure.' OK. In the U.S., these are the critical infrastructure sectors: chemical sector; commercial facilities; communications; critical manufacturing; dams; defense industrial base; emergency services; energy; financial services; food and agriculture; government facilities; healthcare and public health; information technology; nuclear reactors, materials, and waste; transportation systems; and water and wastewater systems. In other words, just about anything humans touch and work with will be impacted by the EU AI Act.

Now, let us look at one critical infrastructure sector: power, which includes electricity, oil, gas, coal, fission, fusion, wind, solar, geothermal, hydropower (dams), biomass, tidal, and biofuels. Let us look at just one. Electricity infrastructure can be broken down to generation, transmission, and distribution. Generators produce electricity along with designed equipment and systems using pressure boilers, water flow, solar (sun heat) collectors, wind, nuclear reactors, pressure vessels, contained / pumped gases, and other means. Transmission systems deliver power for endless applications in industry, business, military, transportation, and our personal lives And more than 50 power systems can generate electricity from wind, hydro, solar, ocean, and biomass.

High-Risk AI

High-risk AI systems have a high likelihood and high consequence of harm to humans. These systems must be safe and reliable. Examples include:

- **Self-driving cars**: Vehicles are designed to operate safely in different conditions as failures lead to accidents.
- **AI in healthcare**: AI systems diagnosing diseases or recommending treatments must be very accurate to avoid harming patients.
- **Facial-recognition technology**: Law enforcement uses AI technology so it is important to prevent misuse and protect privacy.

Limited-Risk AI

Limited-risk AI systems have a lower risk to humans but still require oversight. These systems must meet transparency requirements for humans interacting with AI. While these systems are not as heavily regulated as high-risk AI, they still must be architected, designed, deployed, and assured responsibly. Examples include:

- **Chatbots:** When humans chat with a customer service bot, the organization informs humans of the AI.
- **Recommendation systems:** AI suggests movies or products based on preferences but must disclose how it makes recommendations.

Minimal-Risk AI

Minimal-risk AI systems have little risk to humans. These systems do not require specific regulation. AI systems in this category can be used without oversight or human-in-the-middle oversight. Examples include:

- **Simple games**: AI in video games do not affect humans
- **Virtual assistants**: AI does simple tasks like setting reminders or answering basic questions.

Lesson Learned – EU AI Act Context

The EU AI Act has evidence-based requirements for high-risk AI systems. For example, EU Act articles require:

- **Risk management system:** Article 9 requires high-risk AI providers/developers to implement a risk management system to identify, assess, and mitigate risks throughout the lifecycle of their AI systems
- **Data governance:** Article 10 requires providers/developers to have training, validation, and testing datasets including measures to minimize bias.
- **Technical documentation:** Article 11 requires providers/developers to document compliance with accuracy, robustness, and cybersecurity requirements.
- **Automatic logging:** Article 12 requires high-risk AI systems to include automated logging capabilities for traceability and post-market monitoring.
- **Transparency:** Article 13 requires providers/developers to disclose system functionality, accuracy levels, and human-in-the-middle oversight to allow safe use by AI deployers.
- **Human oversight:** Article 14 requires AI systems to be designed to allow human intervention to prevent human harm or mitigate risks.
- **Accuracy, Robustness, and Cybersecurity**: Article 15 requires high-risk AI systems to meet performance reliability and resilience standards.[78]

Lesson Learned – Focus on Risk

A common thread throughout **Trust Me: AI Governance** is AI technology is running ahead of legislative rule making and governance. Why? Politicians and policy makers are not engineers and do not understand the implications of AI technology. So what are policy makers doing? Kevin Werback, Wharton Professor of Legal Studies, captures this:

> "No country in the world is really seriously trying to regulate all AI. They're focused on use cases or they're focused on risk."[79]

EU AI Act Risk Management Requirements

Article 9 of the EU AI Act states by August 2, 2026 providers/developers of high-risk AI systems must have:

> "risk management system shall be established, implemented, documented and maintained in relation to high-risk AI systems. The risk management system shall be understood as a continuous iterative process planned and run throughout the entire lifecycle of a high-risk AI system, requiring regular systematic review and updating."[80]

Risk Management System Requirements

ISO 31000 should be used to comply with this requirement as well as to develop assurance for high-risk AI systems. Here is a summary of the risk management system requirements for high-risk AI systems:

1. **Risk identification and assessment**
   - Identify known and foreseeable risks to health, safety, and rights.
   - Evaluate risks from both intended use and reasonably foreseeable misuse.
   - Consider risks identified through post-market monitoring.
   - Focus on risks that can be mitigated or eliminated.
2. **Risk management measures**
   - Implement targeted measures to address risks.
   - Balance different requirements to minimize risks effectively.
   - Assure residual risks are acceptable.
   - Prioritize risk mitigation throughout design.
   - Implement risk-control measures for unavoidable risks.
   - Train AI deployers.

3. **Testing requirements**

    - Conduct testing throughout AI lifecycle development.
    - Use predefined metrics and probabilistic thresholds.
    - Include real world testing when appropriate.
    - Assure consistent performance and compliance.

4. **Special considerations**

    - Evaluate impacts on minors and vulnerable groups.
    - Consider deployer expertise and intended use context.
    - Allow integration with existing risk management processes required by other EU laws.

Product Liability Laws

EU product liability laws are designed to protect humans from harm caused by defective products. Product liability laws generally hold manufacturers and sellers responsible if products are unsafe or do not work as intended. As AI becomes integrated into products, product liability laws now apply to AI development and governance. However, this is a contentious issue.

What is Product Liability?

Product liability is the legal responsibility of manufacturers and sellers to produce safe products. If a product causes human injury or damage due to a defect, the organization may be liable for damages. New product liability regulations clarify who is accountable for defects. Proposals suggest AI systems are treated like other products, meaning developers are liable for damages caused by defects, even if those defects come from changes that are made after the product is released.

As AI technology is incorporated into products like self-driving cars, smart appliances, or medical devices, product liability laws may apply:

- **Manufacturing defects:** If an AI system is improperly programmed leading to a malfunction, the manufacturer could be liable. If a factory robot has a software bug causing it to operate dangerously, the designer could be responsible for injuries.
- **Design defects:** If an AI system is designed poorly, it could be a design defect. If healthcare AI makes an incorrect diagnosis due to flawed algorithms, the organization could be liable for resulting harms.
- **Warning labels**: If product lacks user instructions or warnings which could lead to human harm, warning labels are required such as instructions for use (IFUs).

Who is at Fault?

AI systems can also result in cascading and amplifying risks. For example integrated AI agents combine multiple capabilities such as perception, reasoning, learning, and autonomous decision-making. A failure in one interconnected AI component can lead to widespread and compounded impacts.

Determining AI liability and who is responsible when something goes wrong can be difficult. Since AI systems can be embedded in a commercial product, it can be difficult to find fault with the product manufacturer, software developers, or human misuse.

Lesson Learned - AI Liability

We have been working on critical infrastructure products for years. Each week, we see new AI laws and regulations that can impact our assurance app. This means more liability. We limit this through insurance, EULA (end user licensing agreement), consent, and other options. AI agents, AI reasoning, and autonomous decision-making machines are increasing daily. Developers are having a hard time understanding their product liability.

EU Digital Laws

EU is also developing digital privacy, data use, and AI legislation. These laws will be legislatively linked. So, AI governance professionals need to be familiar with them.

DORA

The EU developed the Digital Operational Resilience Act (DORA) to regulate cybersecurity and operational resilience in the financial sector. DORA was adopted in 2022 and went into effect in 2025. The goal is for financial institutions to respond and recover from cyber-attacks and disruptions. The scope includes banks, insurance companies, crypto, and investment firms.

DORA indicates how the EU will tackle AI legislation, regulation, and standardization. DORA requires risk management, incident reporting, third-party risk management, and testing. EU AI Act requires similar AI governance practices, which will become global AI benchmarks.

Lesson Learned – Early Warning Sign

GDPR is an early warning sign or opportunity for AI governance professionals. GDPR also points to the future of AI regulation and governance.

The EU is developing prototype AI legislation. As a first mover EU policy makers face many risks. Abhay Parasnis, CEO of Typeface.ai said:

> "We are seeing a kind of a Wild West situation with AI and regulation right now. The scale at which businesses are adopting AI systems isn't matched by clear guidelines to regulate algorithms and help researchers avoid the pitfalls of bias in datasets. We need to advocate for a system of checks and balances to test AI for bias and fairness and to help businesses determine whether use cases are appropriate for this technology at the moment."[81]

Quality Management System (QMS)

Article 17 of the EU AI Act requires that:

> "providers of high-risk AI systems shall put a quality management system in place that ensures compliance with this Regulation" by August 2, 2026.[82]

QMS Requirements

A quality management system (QMS) is part of the EU's conformity assessment scheme, which will require developers of high-risk AI systems to be certified to ISO 9001 by third-party assessors or possibly self-certify. QMS requires documented evidence in the following areas:

- **Regulatory compliance**: Procedures are developed for conformity assessments and managing system modifications.
- **Design management**: System design, design control, and verification are developed and documented.
- **Quality assurance**: Quality management, assurance, and control systems are deployed and documented.
- **Testing**: Testing, verification, and validation before, during, and after development are documented.
- **Specifications**: Standards for ensuring compliance are deployed and documented.
- **Data governance**: Data acquisition, collection, analysis, storage, and retention are managed and documented before AI deployment.

Lesson Learned – Legislation and Regulation Conflation

Legislation and regulation are often conflated. American legislation consists of laws enacted for example by the U.S. Congress. Regulations are mandatory 'shall' standards and rules enacted by a governmental authority.

AI governance is documented to demonstrate compliance with legislation and regulations that are context-specific. For example, AI governance of a pharmaceutical company in the U.S. must comply with the Food and Drug Administration (FDA) regulations. AI governance of an AI system sold in the EU must comply with AI Act.

EU AI Act Compliance

EU AI Act regulates AI systems so they are safe and ethical. There is an enforcement framework along with penalties for those who do not comply.

Importance of Compliance

Compliance is binary. A company either complies or does not comply. The penalties for noncompliance are designed to encourage companies to take the regulations seriously. By ensuring AI systems are safe and ethical, the EU wants to protect humans. Companies with strong compliance programs want to avoid hefty fines.

Enforcement Framework

In high-risk AI systems there is enforcement with consequences. The EU AI Act prescribes enforcement from:

- **National regulatory bodies**: Each EU member state has a regulatory body responsible for overseeing AI compliance in that country.
- **Compliance checks**: Regulatory authorities conduct audits and checks to meet standards.
- **Reporting mechanisms**: Companies may be required to self-report AI risks or breaches related to AI systems.

Penalties for Non-Compliance

If the AI organization fails to comply with the EU AI Act, it could face penalties. The penalties are tiered based on the risk severity of the violation:

- **Penalties**: If a company engages in banned practices such as using social scoring systems, it faces fines of up to €35 million or 7% of its total worldwide annual turnover, whichever is higher.
- **Fines**: If a company fails to meet transparency requirements or does not properly manage risks, it can be fined up to €10 million or 2% of worldwide turnover depending on which amount is higher.
- **Legal consequences**: Companies can face legal actions, which include lawsuits or restrictions to operate within the EU.

EU Necessity and Proportionality

Necessity is an AI governance principle when assessing fundamental rights such as protection of personal data:

"Necessity shall be justified on the basis of objective evidence and is the first step before assessing the proportionality of the limitation."[83]

"Proportionality is a general principle of EU law. It restricts authorities in the exercise of their powers by requiring them to strike a balance between the means used and the intended aim."[84]

Lesson Learned - AI Governance

As developers, we are considered a provider in the EU AI Act. We must be compliant with the AI Act by August 2, 2027. So, we have time to comply before placing AI systems into the EU market.

The EU regulatory approach is to come up with new regulations, rules, annexes, articles, and interpretations. Specific regulations will be developed for each high-risk AI domain. If these are not done, the EU Act may become window-dressing legislation that is not measurable, actionable, assureable, and enforceable.

Chapter Summary

- AI governance professionals need to understand AI fears:

  "The release of ChatGPT two years ago set off fears that AI could soon be capable of surpassing human intellect and with that capability comes the potential to cause superhuman harm. Could terrorists use an AI system to learn how to build a bioweapon that kills a million people? Could hackers use it to run millions of simultaneous cyber-attacks? Could the AI reprogram and even reproduce itself?"[85]

- AI regulations are coming. Will they be prescriptive, voluntary, or descriptive? This will depend on the culture and AI use case.

- We do know regulations will be voluntary in the U.S.

- Laws prevent algorithmic discrimination by requiring AI systems to treat all humans equally.

- EU AI Act establishes a framework based on the risk level of AI systems.
    - High-risk AI requires risk management, safety, and accountability.
    - Limited-risk AI requires transparency.
    - Minimal-risk AI has the least regulation.
- General purpose AI systems have specific requirements under the EU AI Act so they are safe, transparent, and effective.
- Requirements include risk classification, documentation transparency, data governance, risk management, human oversight, and quality management.
- Non-discrimination laws describe how AI systems should be fair across employment, credit, lending, housing, and insurance sectors.
- Risk classification framework for AI categorizes systems based on risks and impacts.
- EU AI Act establishes different requirements based on the organizational context of AI systems.
    - Providers/developers have extensive responsibilities focusing on development and compliance.
    - Deployers are responsible for how AI is used.
    - Importers are responsible for compliance with EU standards.
    - Distributors can only sell compliant AI systems.
- EU AI Act enforcement framework requires national regulatory bodies to monitor compliance and conduct checks.
- Companies failing to comply face penalties of millions of euros or a percentage of global revenue.

CHAPTER 6

AI STANDARDS & GUIDELINES

What is the Key Idea in this Chapter?

The key idea in this chapter is understanding AI standards and guidelines is critical for AI governance. In this chapter, we discuss the main AI standards and guidelines We also cover different approaches to AI governance. Why do these matter? AI governance professionals are becoming trusted advisors on developing and implementing AI standards and guidelines.

AI Governance Strategies

AI governance professionals serve as trusted advisors to executive leadership, presenting strategic AI governance recommendations, addressing AI risks, and providing guidance on AI best practices. AI governance advisors recommend standards and guidelines that manage risks such as bias, data privacy concerns, or regulatory compliance.

Principles-Based AI Governance

OECD established a set of principles promoting the development and use of trustworthy AI. These principles are a guideline for countries and organizations to develop safe and ethical AI systems.

Principles-Based Governance

There are several regulatory approaches to AI governance. One is the declarative and prescriptive approach using rules and 'shalls' in the legislation, regulations, or standards.

Another form of AI governance is discretionary relying on the company to determine how to comply with a regulation. This approach uses 'should's in the regulation. Both are acceptable approaches for largely static technologies. However, AI is different. AI is dynamic. Static AI governance approaches do not work.

Another approach to AI governance is the principles-approach. Unlike rigid rule-based frameworks, the principles-approach emphasizes flexibility and adaptability to comply and align with the changing technology.

AI Principles – Advantages and Disadvantages

Many companies use a principles-approach for AI governance. There are advantages and disadvantages to this approach. Advantages include flexibility to adapt quickly to AI advances and unforeseen situations. Principles can also be tailored to different contexts and use cases.

The principles approach also has disadvantages. Organizations want precise rules and prescriptive guidance on how to comply with AI statutes and regulations. Principles are often mushy. They lack clarity, which can lead to inconsistent interpretation and application. Principles can be difficult to measure, manage, and enforce. Monitoring and assurance also becomes more difficult.

Google's AI Principles

Google uses an AI principles approach which focuses on the following:

- "Be socially beneficial.
- Avoid creating or reinforcing unfair bias.
- Be built and tested for safety.
- Be accountable to people.

- Incorporate privacy design principles.
- Uphold high standards of scientific excellence.
- Be made available for uses that accord with these principles."[86]

OECD AI Principles

OECD principles focus on:

- **Human-centric approach**: AI is designed to improve human safety and respect human rights.
- **Transparency**: AI systems are transparent meaning that humans understand how these systems work and how decisions are made.
- **Fairness**: AI is developed and used in a fair way.
- **Accountability**: There is accountability for outcomes of AI systems.
- **Safety**: AI systems are reliable and safe.

OECD also suggests its own risk management framework to help countries and organizations deploy principles. This framework includes:

- **Evaluation**: OECD provides guidelines on how to design AI systems to align with trustworthy principles.
- **Collaboration**: OECD emphasizes collaboration among governments and businesses to share best practices.

OECD principles are difficult to put into practice or operationalize. This is a reason why EU and others are contextualizing OECD practices.

Lesson Learned - AI is Exponential – Policy is Glacial

It is hard to keep up with AI. AI change is exponential. Standards development is glacial. So, the delta or gap between AI development and standards development is increasing. Stability AI CEO Emad Mostaque said:

> "We are just trying to race to keep up with the societal impact of all this. And one of the reasons for creating stability was so that we could create some standards … A lot of the standards around the industry right now, I think are incorrect. Like, scrape the whole internet and then just try and tune it back in the preferences. … If I watched all of YouTube, I'd be a bit crazy too."[87]

U.S. AI Executive Order (EO) 14110

On October 30, 2023, President Biden signed Executive Order 14110, which focused on the safe, secure, and trustworthy development and use of AI. It was the first U.S. AI executive order or legislation that set AI governance guidelines and actions for how AI is developed and used while minimizing risks.

What was EO 14110?

The main goal of Executive Order (EO) 14110 was to regulate AI social and technical risks. The White House believed AI could improve healthcare and make everyday tasks easier. The EO was a social and technical mandate to all Federal departments.

Executive Order (EO) Components

The executive order included the following components:

- **Frameworks**: National Institute of Standards and Technology (NIST) is responsible for developing guidelines that align with the president's executive order.
- **Standards**: EO encouraged the creation of standards to regulate AI technologies and be fair and equitable.
- **Policy development**: Federal agencies were tasked with developing policies to govern AI use in different sectors including healthcare and communications.
- **Evaluations**: High-risk AI systems in the U.S. government were going to be assessed and monitored for compliance.
- **Safety**: High-risk AI systems must be safe and secure.

- **Collaboration**: EO emphasized collaboration among stakeholders including government agencies, labor unions, international allies, and social organizations.

The Good and Bad News
The good news. NIST AI RMF became the basis for President Biden's Executive Order to Federal agencies to adopt the standard. And in 2024, OECD and UN adopted the AI guideline.

The bad news. President Trump revoked Biden's Executive Order. This created uncertainty and chaos in AI governance and standards development.

What was the problem? The Biden Administration focused on AI social and civil rights. The Trump Administration wants to focus on AI technical, safety, and homeland security. The challenge for AI governance professionals is that AI is often seen through the prism of politics.

AI Maturity
AI governance, risk management, and compliance (GRC) are often described in terms of social and technical maturity. AI governance can be developed using NIST AI Risk Management Framework or ISO 42001. Low-risk AI systems focus on voluntary compliance and generic systems using ISO 42001. High-risk AI systems require dynamic risk assessments, risk-based auditing, and assurance to NIST AI RMF.

State and Local AI Laws
So, what is happening to AI governance and legislation in the U.S.? We are seeing more state and even local AI regulation. For example, in 2023 the City of New York enacted NY Law 144:

> "A New York City law that comes into effect in January will require companies to conduct audits to assess biases, including along race and gender lines, in the AI systems they use in hiring. Under New York's law, the hiring organization is ultimately liable—and can face fines—for violations." [88]

Future of U.S. AI Regulation

Under President Trump, the U.S. government emphasizes voluntary or self-regulation using the NIST AI RMF. U.S. AI regulation will still be risk-based.

Another trend is that high tech companies are developing their own AI governance frameworks that are based on the EU AI Act and NIST AI RMF.[89]

Lessons Learned - Governance Approaches

President Biden's focused on AI social equity and mandating AI compliance. President Trump's focus is on AI technical safety and voluntary compliance. These are different approaches to AI regulation and governance. The former promotes legislation, conformity, compliance, and enforcement. The latter approach focuses on guidelines, customer reviews, and voluntary compliance.

Peter Diamandis captures what is happening in U.S. AI regulations:

> "If the government regulates against use of drones or stem cells or artificial intelligence, all that means is that the work and the research leave the borders of that country and go someplace else."[90]

NIST AI Risk Management Framework (AI RMF)

The NIST AI Risk Management Framework (AI RMF) is a set of guidelines created by the National Institute of Standards and Technology (NIST) to help organizations manage AI risks. NIST also developed a Playbook that provides practical steps for implementing these guidelines. Google 'NIST AI RMF' and you will get the framework and playbook for free.

Core Functions of the AI RMF

The AI RMF is built around four core functions that guide organizations in managing AI risks:

- **Govern**: Governance includes oversight and monitoring policies and procedures.

- **Map**: Mapping involves recognizing and understanding AI system risks. Organizations assess what risks come from AI technologies including ethical concerns, safety issues, biases, and security vulnerabilities.
- **Measure**: Measure includes evaluating the consequence and likelihood of possible risks. The assessment also prioritizes risks that need monitoring.
- **Manage**: Manage focuses on implementing strategies to mitigate the identified risks. Organizations develop plans to address the risks, which include improving the AI system, changing how it is used, or providing additional training for humans.

The above AI RMF core functions are illustrated on the next page.

Categories and Subcategories

The AI RMF is organized into categories that offer guidance on specific elements of the AI risk management framework:

- **Categories**: Are broad areas of the AI risk management framework.
- **Technical**: Focuses on the technical elements of AI systems such as algorithms, data quality, and system performance.
- **Ethical**: Addresses ethics, including fairness, accountability, and transparency in AI decision-making.
- **Security**: Includes protecting AI systems from threats and vulnerabilities, assuring they are secure from attacks or misuse.
- **Societal**: Looks at the broader human impacts of AI on society.

NIST AI RMF Advantages and Challenges

The NIST AI RMF has advantages and challenges. Let us first look at the advantages:

- **AI governance:** It is the foundation document for U.S. and possibly global AI governance, responsible policies, and ethical practices.
- **Manage governance:** It helps organizations implement and future proof responsible governance throughout the AI lifecycle.

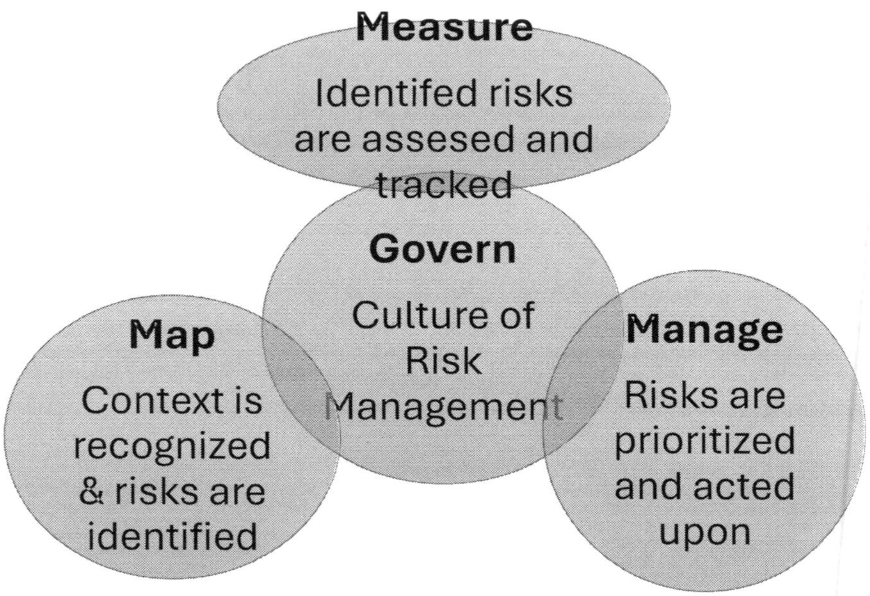

- **ERM system integration:** It is an AI ERM guideline. ERM is a top down, strategic view of AI risk management.
- **Risk management guidelines:** It has comprehensive guidelines to identify, assess, and mitigate AI risks of simple to highly complex AI systems.
- **Guideline integration:** NIST AI RMF, NIST CSF 2, and NIST Privacy Framework are risk-based. NIST will integrate these guidelines and subsume them under an ERM umbrella.
- **Social and technical guidelines:** It addresses social and technical challenges in the design, deployment and testing of AI systems throughout the lifecycle.
- **Application:** It is applicable to different industries, use cases, contexts, and environments. AI RMF will increasingly be integrated into sectors like healthcare, finance, critical infrastructure, and national defense.
- **It is free:** NIST AI RMF and NIST AI RMF Playbook can be downloaded for free at: https://www.nist.gov/itl/ai-risk-management-framework. ISO 42001 and IEEE AI standards are expensive.

- **Best practices guideline:** It was developed by a U.S. Federal agency and will become the de facto guideline for AI standards and guidelines.
- **Legal standard:** It is the foundation document for emerging AI regulations in the U.S., such as Colorado's Consumer Protections for Artificial Intelligence law.
- **Legal defense:** NIST AI RMF compliance has a legal defense and cover in some jurisdictions as the AI standard of due care similar to what NIST Cybersecurity Framework 2.0 is for cybersecurity.

NIST AI RMF Challenges

The NIST AI RMF including Playbook face a number of challenges:

- **AI systems are complex:** AI systems are inherently complex making it difficult to identify, measure, and manage risks effectively, efficiently, and economically.
- **AI RMF is a guideline not a standard:** As a guideline, NIST AI RMF means that a company has the option of whether to use it or not.
- **Resource dense:** NIST AI RMF requires deep technical resources for complex and high-risk AI systems.
- **Is descriptive, not prescriptive:** NIST AI RMF is descriptive so it can be used in different contexts. The challenge is the language in the guideline can be interpreted in different ways.
- **Political document:** NIST AI RMF was developed under the liberal Biden administration. AI RMF addresses social and technical AI use cases. The challenge is the Trump administration focuses on the security and technical elements of AI.
- **Difficult to operationalize:** NIST AI RMF with its Playbook are massive. A novice developer, user, or stakeholder will not know how to enter, use, and optimize the framework.

- **AI challenges:** NIST AI RMF can be used with different AI technologies. This can be difficult for emerging technologies like generative AI and large language models (LLMs) where risks evolve rapidly.
- **AI risk management challenges:** NIST AI RMF does not illustrate how to conduct different types of AI risk assessments. To remedy this, NIST is developing use cases.

AI Safety Institutes

As part of NIST, the U.S. set up the first AI Safety Institute (AISI) to evaluate AI safety. Australia, Canada, EU, France, Japan, Kenya, Republic of Korea, Singapore, and the United Kingdom then developed their own AI Safety Institutes. According to the Mission Statement of the International Network of AI Safety Institutes:

> "Our institutes and offices are technical organisations that aim to advance AI safety, help governments and society understand the risks posed by advanced AI systems, and suggest solutions to address those risks in order to minimize harm. Beyond mitigating risks, these institutes and offices play a crucial role in guiding the responsible development and deployment of AI systems."[91]

AISIs now are being politicized. The UK renamed its AISI to AI Security Institute. Trump administration may defund or close the U.S. AISI.

Lesson Learned – Be Nimble & Be Quick

U.S. President Biden mandated federal agencies use and deploy the NIST AI RMF. This was great. We thought we had the perfect evergreen app. It was mandated. Then, Trump administration revoked Biden administration AI mandates. As a developer, we focused on Biden's AI RMF requirements but with President Trump, we had to pivot our apps to EU AI Act requirements. Our lesson learned: Be nimble. Be quick.

NIST ARIA Program

NIST developed the ARIA Program, which stands for **A**ssessing **R**isks and **I**mpacts of **A**I. NIST AI governance resources can be found at https://ai-challenges.nist.gov/aria.

Purpose of the ARIA Program

The ARIA program helps organizations and humans evaluate AI technologies. The program develops structured approaches and processes for assessing the risks of AI systems.

ARIA provides the following guidelines for organizations:

- **Assessment frameworks**: Structured guides help organizations systematically evaluate AI systems.
- **Testing protocols**: Specific procedures are used to test AI systems.

Metrics and Measurements

The ARIA program also helps the U.S. AISI promote safe AI practices. These criteria quantify AI system safety and performance for:

- **Safety metrics**: Measurements assess whether the AI system operates without causing harm.
- **Performance metrics**: Criteria evaluate the accuracy and efficiency of the AI system.
- **Fairness metrics**: Measurements check for AI system biases.

Lesson Learned – Where You Sit is What You See

AI governance professionals need to understand the difference between government and business requirements. Government addresses public policy problems. Entrepreneurs solve business problems. This makes a big difference in AI governance. Governments and policy makers want to solve societal problems such as equity, trust, and social stagnation. On the other hand, developers and engineers want to develop and marketable products. Another way to express this, governments want to regulate. Developers want to productize and monetize.

HUDERIA and New Tools

New AI standards and guidelines will be developed over the next several years. HUDERIA is an example of a new AI safety standard. A little history. In 2025, the Council of Europe developed the Methodology for The Risk and Impact Assessment of Artificial Intelligence Systems from The Point of View of Human Rights, Democracy And The Rule Of Law. This mouthful is also known as the HUDERIA Methodology.

What is HUDERIA?

HUDERIA is a social approach for evaluating the risks and impacts of AI systems on human rights, democracy, and rule of law. The Council developed a risk and evidence-based approach to identifying, preventing and managing harms to humans throughout the AI lifecycle.

Another reason why the Council developed HUDERIA was to be able to assign accountability if things went wrong. HUDERIA is free. Google 'HUDERIA and AI' and you will get the framework.

Again, what a human sees depends on where he or she sits. Developers saw this as regulatory overreach for the Council to establish blame and punish perpetrators through accountability litigation.

HUDERIA Components

HUDERIA is composed of:

- **Context-Based, Risk Analysis (COBRA):** This identifies AI risks based on the social and technical context of the AI system.
- **Stakeholder Engagement Process (SEP):** This identifies stakeholders who could be harmed by the AI system.
- **Risk and Impact Assessment (RIA):** This evaluates risks to human rights and the rule of law.

- **Iterative preassessments:** This assures AI systems are safe throughout their lifecycle through continuous monitoring and auditing.

Why HUDERIA?

HUDERIA was developed to bridge the social and technical elements of AI governance and risk management. As the Council said, HUDERIA is a guideline and not legally binding. From a public policy perspective, there are constraints and gaps in AI regulation. The purpose of HUDERIA is to fill the gaps.

Lesson Learned – Europe as an AI Museum

I was watching Davos and AI House programs on YouTube. The EU was very proud taking the global lead on AI regulation. Some were warning that excessive regulation would stifle innovation. And it was noticeable that the EU was falling behind in AI. One of the AI panelists quipped if the EU bureaucrats stifled innovation, the EU could be relegated to an AI museum.

ISO 42001 AI Management System (AIMS)

The International Organization for Standardization (ISO) develops standards to guide the development of AI. Two important AI standards are ISO/IEC 22989 and ISO/IEC 42001. IEC is the International Electrotechnical Commission, which like ISO is based in Geneva, Switzerland.

ISO 22989: AI Concepts and Terminology

ISO 22989 provides taxonomy guidance for AI systems' development. If you are interested in understanding the standard in more detail, purchase **Trust Me: Risk Management** from Amazon. The standard covers:

- **Terminology**: Standard defines terms and concepts related to AI.

- **Standards**: ISO 22989 supports the development of other ISO AI standards.

- **Taxonomy**: AI taxonomy establishes a common platform to set up an AI governance program.

ISO 42001: AI Management Systems

ISO 42001 is the first AI Management System (AIMS) standard. The standard may be used as part of the EU AI Act conformity assessment scheme. The standard may be used for first, second, and third-party assessment to validate compliance.

A full review and explanation of ISO 42001 is beyond the scope of this book. You may want to purchase **Trust Me: ISO 42001** from Amazon. ISO 42001 focuses on:

- **AI Management Systems (AIMS):** The standard outlines the requirements for developing an AI management system.
- **Integration with organizational processes:** The standard describes how to integrate AI management into existing organizational processes.
- **Continuous improvement**: The standard focuses organizations to continuously improve AI systems.
- **Voluntary:** ISO 42001 is a voluntary standard meaning organizations are not legally required to follow it. Yet!!! EU AI Act and other legislation may require ISO 42001 management certification.

ISO 42001 Benefits

The adoption of the AIMS has been slow. ISO 42001 certification provides a low level of AI assurance. It will be used for auditing and certifying low-risk AI systems. ISO identified the following AIMS benefits:

> "ISO/IEC 42001 exists to help businesses and society-at-large safely and efficiently derive the maximum value from their use of AI. Users can benefit in numerous ways:

- "Improved quality, security, traceability, transparency and reliability of AI systems.
- Enhanced efficiency and AI risk assessments.
- Greater confidence in AI systems.
- Reduced costs of AI development.

- Better regulatory compliance through specific controls, audit schemes and guidance that are consistent with emerging laws and regulations."[92]

7 Years to Develop ISO Standard

ISO is a global standards development organization. It develops manufacturing (ISO 9001), information security (ISO 27001), and AI management system standards (ISO 42001). These management system standards are part of the conformity assessment scheme the EU is proposing for AI governance.

The challenge is standards development requires consensus from teams of experts from different countries. Consensus building and developing ISO standards take up to seven years from proposal to final standard. On the other hand, AI development is exponential.

What Framework to Choose?

Choosing and tailoring the appropriate risk management framework is a major decision. The appropriate framework depends on an organization's context, size, industry, and specific needs. Once, the AI risk management is chosen, then it needs to be tailored to the AI use case.

On the next page, five risk management frameworks are evaluated: 1. NIST AI RMF; 2. EU AI Act; 3. ISO 31000; 4. ISO 42001; and 5. HUDERIA. It is important to choose the right framework. ISO 42001 is a 10 page standard. The other 40 or so pages describe controls and definitions. So, we call ISO 42001 a light AI guideline which may not be appropriate for high-risk AI systems. These systems would require an ERM system like ISO 31000, which is covered in the next chapter. It is a risk management framework that describes a generic risk taxonomy, processes, metrics, and outcomes that can be tailored for high-risk AI systems.

Lesson Learned - Do Not Boil the Ocean

Quick story: Our company was retained to help a company get certified to ISO 27001, ISO's Information Security Management Standard (ISMS). The customer had dropped $250,000 with another consulting firm, who could not complete the consulting project. The previous company was fired. We were retained to fix the problems and get the company certified.

Aspect	NIST AI RMF	EU AI Act	ISO 31000	HUDERIA	ISO 42001
Primary Focus	Risk management for AI systems, emphasizing trustworthiness and reliability.	Regulation of AI systems, focusing on risk-based classification and compliance.	General risk management principles applicable across industries.	AI systems' impact on human rights, democracy, and rule of law.	Management system standard for AI, focusing on responsible development.
Scope	Broad, covering all AI systems and their lifecycle.	Specific to AI systems within the EU, with a focus on high-risk applications.	Universal, applicable to any organization or sector.	Focused on societal and ethical impacts of AI systems.	Applicable to organizations involved in any part of the lifecycle.
Risk Approach	Structured risk management framework tailored for AI systems.	Risk-based classification (unacceptable, high, limited, minimal risk).	General risk management principles (identify, assess, treat, monitor risks).	Focuses on risks to human rights, democracy, and rule of law.	Establishes a framework for i managing risks and opportunities
Legal Binding	Voluntary framework for organizations.	Legally binding regulation for EU member states.	Voluntary international standard.	Voluntary assessment framework.	Voluntary international standard.
Target Audience	AI developers, deployers, and organizations using AI systems.	AI providers, deployers, and users within the EU.	Organizations across all sectors and industries.	Policymakers, AI developers, and organizations	Organizations developing, or using AI systems,
Ethical Considerations	Indirectly addressed through trustworthiness and fairness principles.	Explicitly addresses ethical concerns, especially for high-risk AI systems.	Not explicitly focused on ethics; focuses on risk management.	Central focus on ethical impacts, human rights, and democratic values.	Ethical AI considerations including fairness, transparency.
Human Rights Focus	Limited; primarily focused on technical and operational risks.	Indirectly addressed through high-risk requirements.	Not explicitly addressed.	Core focus on human rights impacts of AI systems.	Addresses human rights indirectly
Democracy &	Not explicitly addressed.	Addressed indirectly through transparency	Not explicitly addressed.	Central focus on democracy and rule of law	Addresses indirectly through requirements

Rule of Law		and accountability requirements.		implications of AI systems.	for transparency.
Lifecycle Coverage	Covers the entire AI lifecycle (design, development, deployment, monitoring).	Focuses on pre-market and post-market compliance for AI systems.	Applies to all stages of organizational processes.	Focuses on the societal impacts of AI systems throughout their lifecycle.	Covers AI lifecycle, from design and development.
Transparency Requirements	Emphasizes transparency in AI system design and risk management.	Mandates transparency for high-risk AI systems (e.g., explainability).	Encourages transparency in risk management processes.	Emphasizes transparency in assessing societal and ethical impacts.	Requires transparency in AI system development.
Accountability	Encourages accountability through risk management practices.	Requires accountability for high-risk AI systems (e.g., documentation, audits).	Encourages accountability in risk management.	Focuses on accountability for human rights and democratic impacts.	Establishes clear lines of accountability for AI risks.
Global Applicability	Designed for global use but tailored to U.S. contexts.	Primarily applicable within the EU, but may influence global standards.	Globally applicable across industries and regions.	Designed for global use, with a focus on universal human rights principles.	Designed for global applicability, providing framework for responsible AI management.
Enforcement Mechanism	Voluntary adoption; no enforcement mechanism.	Legal enforcement through EU member states (fines, penalties).	Voluntary adoption; no enforcement mechanism.	Voluntary adoption; no enforcement mechanism.	Voluntary adoption; certification and audits by accredited bodies.
Stakeholder Involvement	Encourages multi-stakeholder collaboration in risk management.	Requires consultation with stakeholders for high-risk AI systems.	Encourages stakeholder involvement in risk management.	Emphasizes stakeholder engagement, especially affected communities.	Requires stakeholder engagement in development and deployment of AI systems.
Technical vs. Societal Risks	Primarily focuses on technical and operational risks.	Balances technical risks with societal and ethical concerns.	Focuses on organizational risks, including technical and operational.	Primarily focuses on societal and ethical risks.	Balance technical, operational, ethical, and societal risks.

We noticed the previous consulting firm had told the client to find cybersecurity risks, put them into the risk register and then develop controls. When we were retained, there were over 10,000 risks with more than 30,000 controls in the risk register. Way too many! No wonder the client was unhappy and looking at a $1,000,000 engagement.

Our solution was straightforward. Focus on the critical few, strategic risks – not the insignificant many. Develop the requisite controls based on the client's risk appetite and use cases. Then, get certified. The company did as suggested. The company got its ISMS registered. Everyone was happy.

Do not try to boil the ocean. Focus on the critical few, not insignificant many AI risks.

Chapter Summary

- The AI governance challenge over the next few years is summarized by *Axios Magazine*:

  "The next 3-4 years will reveal winners and losers in the business, but they will set social customs and workplace expectations around AI: How, where and when we use it; what's OK and what's unacceptable to ask AI to do; and what kinds of haywire responses demand fixes from the AI's owners and makers."[93]

- AI standards cannot keep up with the technology. Jennifer King and Caroline Meinhardt at Stanford University distilled the challenge:

  "AI experts universally agree that there needs to be more discussion and collaboration around AI ethics, privacy, and government regulation. These will continue to be areas that change as technology advances and regulations play catch up."[94]

- AI laws are developed and deployed by countries and organization.

- AI governance can be virtue signaling, like green washing for the environment. It sounds good. It looks good. But so what?

- Executive Order 14110 was important to ensuring AI is developed and used in a safe, secure, and trustworthy manner.
- President Trump rescinded President Biden's EO.
- States in the U.S. are incorporating by reference NIST AI RMF and ISO 42001 in their legislation.
- NIST AI Risk Management Framework and Playbook have a broad approach to managing AI technology risk.
- NIST AI RMF helps organizations build safe, ethical, and trustworthy systems.
- OECD principles are a broad framework for trustworthy AI.
- ISO 22989 provides language and foundational concepts for AI.
- ISO/IEC 42001 focuses on the management of AI systems throughout the AI lifecycle.
- NIST ARIA program is designed to develop safe, reliable, and fair AI systems.

CHAPTER 7
ISO 31000:
ENTERPRISE RISK MANAGEMENT

What is the Key Idea in this Chapter?

The key idea in this chapter is ISO 31000 will become the ERM solution for AI governance of high-risk AI systems. Why is this important? Risk management is the core lens for architecting and managing governance in AI legislation and regulation. AI governance professionals must be able to tailor risk frameworks such as ISO 31000 and ISO 42001 to satisfy AI legislative and regulatory requirements.

AI Governance Strategies

AI governance professionals develop risk frameworks to manage risks, track the performance of AI initiatives, and assure measurable outcomes are aligned with expected ROI. These risk management frameworks also align business and AI outcomes with expected ROI and strategic objectives. ISO 31000 is an ERM approach to mitigate risks of AI systems deployment.

ISO 31000 Importance

The principles of risk management anchor AI governance. Global legislation, regulation, and standards are risk-based. This is the reason why there is a separate chapter on ISO 31000 in an AI governance book.

Why is ISO 31000 ERM Important?

ISO 31000 is important for AI governance professionals because:

- OECD identified ISO 31000 as one of the key sources for AI governance.
- EU AI Act requires a risk management system.
- EU AI Act and global AI regulation require evidence-based, risk management.
- Publicly held companies are required to have an ERM system as a listing requirement.
- ISO 31000 is the global risk management standard.
- ISO 31000 is the foundation for AI governance, risk management, and compliance (GRC).
- Organizations build an ERM framework to proactively, preventively, predictably, and preemptively manage AI risks.
- ISO 31000 is a structured and globally accepted, risk approach that fosters trust among stakeholders.

ISO 31000 Risk Management Principles

ISO stands for International Organization for Standardization. ISO is the largest standards development organization in the world. ISO is represented by more than 119 countries around the world. ISO is composed of standards organizations, which are called national standards bodies.

ISO 31000 Principles

The overarching goal of ERM is to create, maintain, increase, and protect AI value. This is shown in the figure on the next page. Value has several dimensions and elements. Value is the ability to meet organizational goals and objectives. Value includes

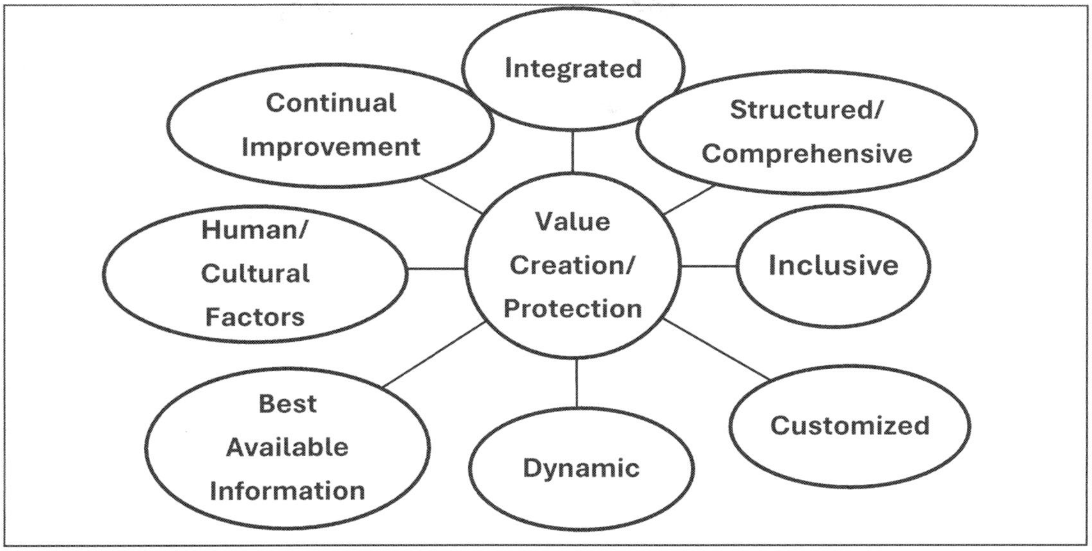

maximizing AI opportunities where upside risks can be exploited such as through new AI development or purchasing AI capabilities. The eight ISO 31000 principles are:

1. **Integrated:** AI risk management is part of AI problem solving and decision-making within the organization.
2. **Structure and comprehensive:** AI risk management is consistent and logical so that AI outcomes are known and AI objectives are achieved.
3. **Customized:** AI risk management framework and processes are tailored to the organizational context.
4. **Inclusive:** AI risk management includes the opinions of critical stakeholders and interested parties.
5. **Dynamic:** AI risk management considers emerging or changing risks due to uncertainty based on the organization's context and environment.
6. **Best available information:** AI risk management is based on actionable and current information, so accurate decisions can be made.
7. **Human and cultural factors:** AI risk management considers cultural and behavioral factors at the following levels: 1. Enterprise; 2. Programmatic/Project/Process level; 3. Transactional/Product level.

8. **Continual improvement:** AI risk management is continuously enhanced through AI lessons learned and experiences that are fed back into the risk management system.

As you read these ISO 31000 principles, map them to the AI regulations, standards, and guidelines with which you have to comply. You will see how ISO 31000 can be used to architect, design, deploy, and assure AI governance.

But Certification Is Not Allowed!

ISO 31000 is an ERM framework. Many publicly registered companies are required to have ERM as part of being listed in an exchange such as the London or New York Stock Exchanges.

ISO 31000 is a guideline that is based on principles. Here is the catch. The standard specifically prohibits it from being certified as an ISO management system standard. Or expressed another way, the standard cannot be used for conformity assessment or certification.

The global market for risk assurance is increasing. Companies want to be risk assured and certified. The marketplace hates a vacuum. If there is a need, there is a way. So, the large global certification bodies are offering ISO 31000 Certificates of Conformance instead of risk management certifications.[95]

Lesson Learned - Risk Tolerance

Risk is the lens for designing, deploying and assuring AI governance. Risk tolerance becomes a critical decision factor for determining the type of AI governance based on AI use cases such as the EU Act's 8 high-risk AI systems.

Over the next 5 years, risk-controls and human oversight through AI governance will be a requisite for organizations and developers. AI governance and ERM approaches will be differentiated and evaluated in terms of use case, application, context, and functionality. For example, AI governance for EU AI Act's high-risk AI systems will be different than that of a chatbot.

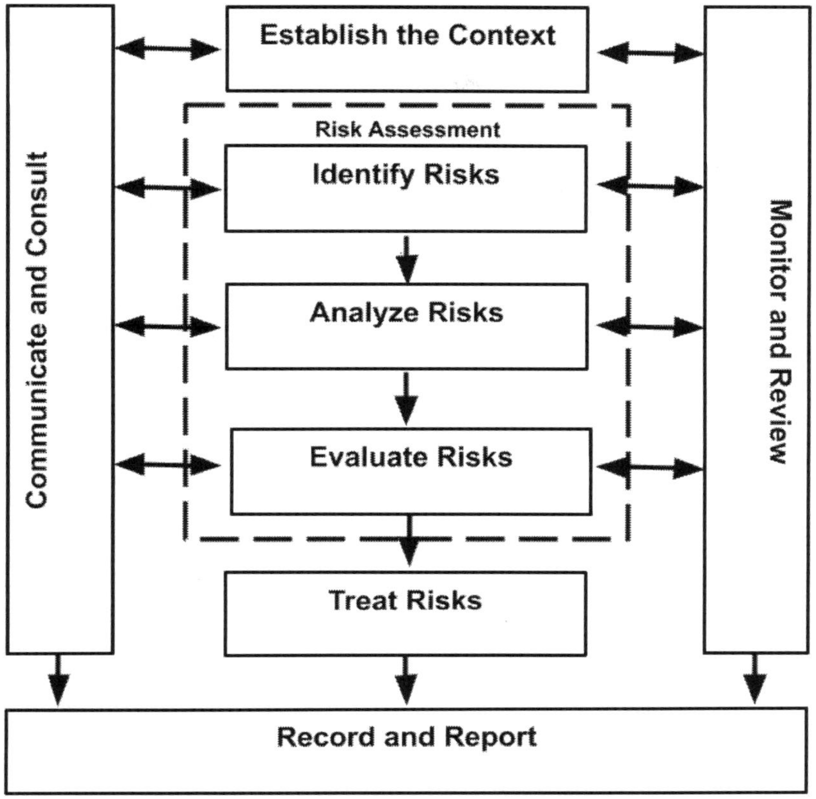

ISO 31000 Risk Management Framework

ISO 31000 can be tailored to a use case because the guideline is a flexible framework rather than a rigid set of instructions. AI effectiveness will be measured in terms of how ERM is integrated into an organization's AI governance and compliance processes.

ERM Governance

What is ERM Governance? The Board of Directors establishes the governance of the organization. Executive management carries out the responsibilities of good governance, risk management, and compliance following AI statutes and rules.

The ISO 31000 risk management process is integral to risk-based, problem solving and decision-making. You can see the ISO 31000 process in the above figure. It looks complicated but is straightforward when reduced to basics. Let us look at each element of the risk management framework.

Communicate And Consult

Communicate and Consult of ISO 31000 ensures stakeholder and interested party AI requirements, needs, and expectations are understood. Stakeholders and interested parties can then make informed risk decisions within the organization.

Internal and external stakeholders understand the AI governance framework, including risks and mitigation strategies. Communication and Consultation facilitates awareness and understanding of risks throughout the organization, supply chain, and among stakeholders. Once critical stakeholders' needs, requirements, and expectations are understood, they can be met.

Establish The Context

AI context defines the goals, scope, boundaries, guard rails, and stakeholders of AI governance such as ensuring fairness, accountability, transparency, and compliance with ethical and legal standards. AI context is determined through use cases. AI context then can define the scope of the AI governance. The idea is not to boil the ocean.

Identify Risks

AI risks can be social and technical risks. AI social risks include: bias, privacy violations, deep fakes, data privacy violations, and unintended behaviors. AI technical risks include cybersecurity threats, autonomous weapons, and high energy use.

AI risks can be identified at each stage of the AI lifecycle – architecting, designing, deploying, and assuring high-risk AI systems. AI risks can include:

- **External risks**: These risks come from outside the organization.
- **Internal risks**: These risks come from within the organization or the AI development process.
- **Data quality risks**: If the data is inaccurate or biased, it can lead to poor performance.

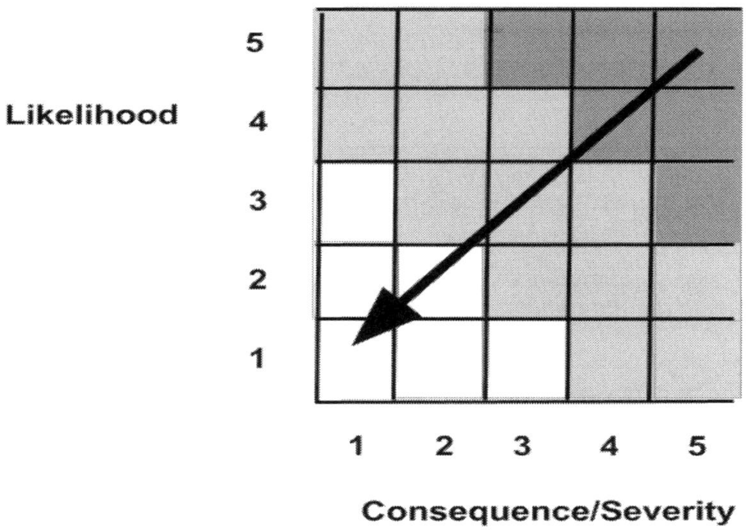

- **Technical failures**: Problems with the AI's algorithms or software can cause a technical malfunction.
- **Regulatory changes**: New laws or regulations such as the EU AI Act impact how the AI is used.
- **Public perception risks**: Negative public opinion of AI affects its acceptance and use.

Analyze Risks

Risk is often analyzed in terms of its likelihood and consequence of harming a human. This analysis results in a heat map such as the one above. Risk analysis attempts to understand the extent, characteristics, and nature of the AI risks. Risk analysis also results in a review of uncertainties, risk sources, events, consequences, probabilities, scenarios, controls, location of risk, quality of information, reliability of information, and scenarios.

Risk analysis can be quantitative or qualitative. The level and extent of AI risk analysis are based on AI context and use cases. Risk analysis depends on the purpose of the analysis, customers, quality information, reliability of information and resources available.

Treat Risks

AI risk treatment is the same as risk management. Risk treatment establishes risk-control processes for monitoring AI systems including automated alerts for anomalous behaviors as defined in the EU AI Act.

To manage AI risks, humans should follow a risk mitigation strategy based on use case and context. The following are 4 ERM treatment strategies:

- **Avoidance**: This includes plans to avoid AI risk entirely. For example, if a data source is known to be biased, developers will not use it.
- **Reduction**: This includes AI risk-controls to reduce the likelihood and consequence of the risk by improving data collection methods or enhancing system testing. Risk reduction is also called risk-control.
- **Transfer**: This means shifting or sharing AI risks to another party such as through insurance or outsourcing AI development.
- **Acceptance**: Humans decide to accept the AI risk if it is low risk, explainable, and manageable.

It is important to note that treatment strategies are based on the risk appetite of the organization. The residual risk is compared against the organization's risk tolerance and appetite. If the residual risk is greater than the risk appetite, then additional treatment is applied to the AI system. This is shown on the figure on the next page.

Monitor And Review

Monitoring and Review gathers insights from stakeholders to refine the AI governance framework. Monitoring and Review defines Key Performance Indicators (KPIs to measure the effectiveness of AI risk treatment including AI accuracy rates, false positives/negatives, explainability and fairness metrics. The AI governance framework is updated to align with evolving AI technologies and regulatory landscapes.

Record and Report

Record and Report is the last element of ISO 31000. Think of Record and Report as documenting key information and then ensuring the right AI information is sent to the right AI stakeholders at the right time.

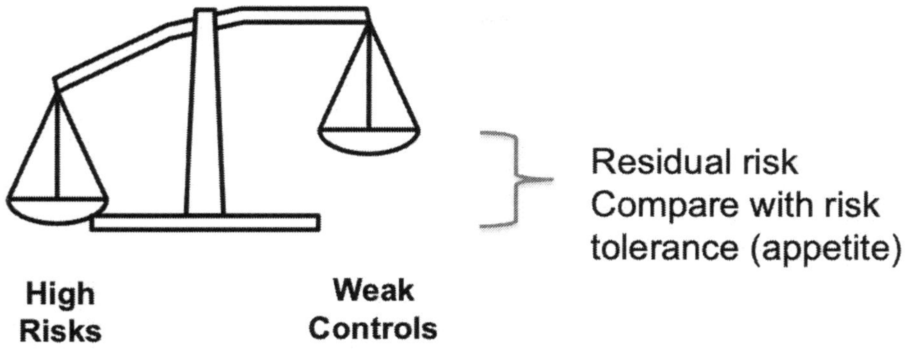

Lesson Learned - Importance of Context

A boss once told me: 'context is worth 20 IQ points.' OK. Now, I would add that 'context is worth 20 IQ (Intelligence Quotient) and Emotional Quotient (EQ) points.' When you read AI legislation and regulations, you will see the term 'context.' AI context defines the scope and type of rules of engagement between humans and machine. Or in other words, context defines how AI governance is architected, designed, deployed, and assured.

Chapter Summary

- NIST and other governmental agencies are subsuming AI, cybersecurity, and privacy guidelines under an ERM umbrella.
- If a company has a high-risk AI system, the company will want to deploy ISO 31000.
- ISO 31000 is referenced by OECD as a 'must use' standard. The challenge is how, when, and who should architect, design, deploy, and assure AI governance.
- ISO 31000 is not an AI specific lifecycle framework. However, it enhances governance and ERM practices within each stage of the AI lifecycle.

- ISO 31000 process has the following elements: communication and consult; establish the context; identify risks; analyze risks; evaluate risks; treat risks; record and report; and monitor and review.

CHAPTER 8
ARCHITECTING AI SYSTEMS

What is the Key Idea in this Chapter?

The key idea in this chapter is AI governance must be architected and tailored into AI systems to identify and mitigate risks and biases inherent in these systems. AI architecture involves AI policies, procedures, and best practices that are baked into the AI system.

What does this mean? AI governance professionals are integral to architecting safe and trustworthy AI systems. By understanding the AI development process, AI governance professionals can reduce data bias, algorithmic fairness, and unintended risks.

Organization of Chapters

Architecting AI systems involves developing use cases and analyzing business processes to discover the risk impacts of AI systems. This architecting process evaluates AI feasibility, legislative AI requirements, strategic alignment, and potential ROI to prioritize high-value projects, uncover inefficiencies, automate repetitive tasks, and optimize workflows.

AI Development Chapters

The next four chapters cover the basics of AI systems development using the architect, design, deploy, and assure process shown in the figure on the next page. These chapters are written from the perspective of the governance professional and AI developer:

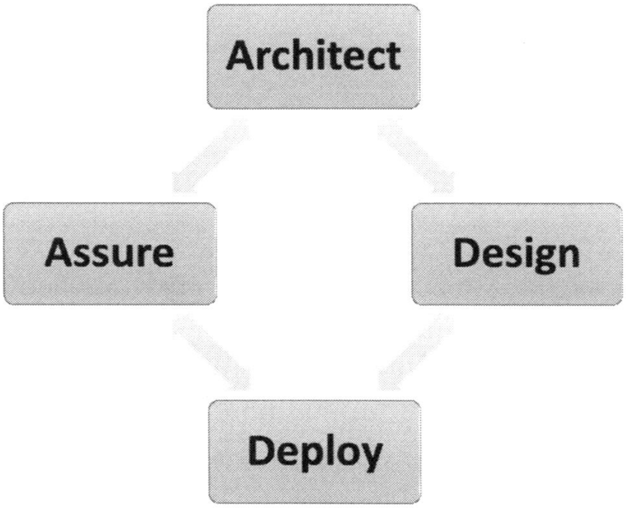

- **Chapter 8: Architecting AI Systems:** The key idea in this chapter is AI governance must be architected and tailored into AI systems to identify and mitigate risks and biases inherent in these systems.
- **Chapter 9: Designing AI Systems:** The key idea in this chapter is understanding how AI systems are designed and built is critical for effective AI governance.
- **Chapter 10: Deploying AI Systems:** The key idea in this chapter is understanding the critical steps of moving an AI system into a real-world environment helps to develop realistic AI governance.
- **Chapter 11: Assuring AI Systems:** The key idea in this chapter is assurance, monitoring, and testing are key to building trust in AI systems. AI assurance is the essence of AI trust.

In the chapter, we often refer to 'AI systems' and 'AI models.' In critical infrastructure such as a water treatment plant, an AI system includes hardware controls, software controls, databases, and user interfaces. An AI model such as an algorithm is usually a single component of a larger AI control system.

AI System Developer Stories

The material in the next 4 chapters are guidelines. They are not standards or prescriptions. The architecture, design, deployment and assurance of AI systems should be tailored to the use case, context, and risk of the AI system. A high-risk AI system in a chemical plant would be different than one in a mindfulness app.

AI Governance Points of View

AI systems development is based on context. AI systems development has stakeholders with different interests and different points of view because of different priorities, agendas, and goals. For example, governments want to focus on autonomous decision-making in national security and critical infrastructure. Companies want ROI and less safety regulation. Regulatory bodies want enforceable legal frameworks and standards. AI developers want rapid innovation and revenue with minimum regulation. Users want easy to use, fair, and explainable AI systems.

In the following chapters, we discuss AI governance from various points of view, competing interests, and various risks among the different stakeholders.

Lesson Learned – ADDA Tailored Governance Framework

Critical infrastructure protection (CIP) is among the highest risk AI systems. Quality Plus Engineering has been evaluating these for over 20 years. We developed the Architect – Design - Deploy – Assure® tailored process for these high-risk systems. Each company or developer can do the same.

Several examples illustrate tailoring AI processes. IBM developed the AI system Lifecycle Management framework. Groove Technology structured an AI Development Lifecycle with seven stages. The point is every governance framework must be tailored to the AI context and use case. So, take the information in **Trust Me: AI Governance** to develop your own AI governance solution.

AI Governance Product Manager

AI professionals need to know how AI models are built. Why? AI governance has to be architected and designed into the AI system not bolted on after the system it deployed. We have seen AI governance rookies take AI governance manuals and simply change company firm names. Or, AI governance rookies may ask ChatGPT to architect an AI manual.

AI governance professionals have to evolve into AI governance product managers. This combines traditional product management responsibilities with a specialized focus on risk frameworks such as ISO 31000 and ethical AI practices such as NIST AI RMF. The AI governance Product Manager can then become a value add instead of friction and cost center.

Lesson Learned – Joint Approach

Many AI governance professionals have never developed an AI pp. This begs the question: How can a person govern and regulate an area where he or she has no work experience?

We advocate for a joint approach to AI governance involving Product Management, Legal, Engineering, and Compliance. Each discipline offers value. However, value will be synergized in the combined approach. McKinsey, the strategy consulting firm, expressed it as:

> "Capturing value from Gen AI will require a concerted, cross functional approach that includes regular input from across the organization, starting with the Chief Executive Officer (CEO). Forward looking enterprises will want to consider altering how they price and package their products from quantity of humans to usage, as well as the possibility of new tiers for Gen AI features."[95]

AI Legislation and Regulation

When developing or using an AI system, developers and AI governance professionals must understand AI legislation and regulations. As we have discussed, U.S. state legislation references NIST AI RMF and ISO 42001.

AI Related Laws

In the U.S., NIST privacy, cybersecurity, and AI guidelines are based on risk management and are integrated into the ERM approach. Since AI relies on large amounts of data, including personal information, AI governance professionals need to know:

- **State privacy laws**: In the U.S., states have enacted their own privacy laws such as the California Consumer Privacy Act, which gives residents rights regarding personal information.
- **Anti-discrimination laws**: AI systems can unintentionally discriminate against groups of humans. Laws prevent unfair treatment based on characteristics like race, gender, or age.
- **Equal opportunity laws:** Hiring practices must be fair and do not discriminate against applicants based on protected characteristics. If an AI system is used in hiring, it must comply with these laws to avoid bias.
- **Consumer protection laws:** These laws are designed to protect consumers from unfair or deceptive practices.
- **Truth in advertising:** AI generated content should not mislead consumers. If an AI system creates ads, the ad is not deceptive.
- **Intellectual property laws:** When developing AI systems, intellectual property laws impact AI governance. Intellectual property includes copyrights, trademarks, patents and trade secrets.

Sector Specific Regulations

We are now seeing two types of AI legislation: 1. Broad protection of AI safety and rights and 2. Industry-specific rules and regulation. Let us look at each type. As AI technology evolves, new laws and regulations are developed to address its unique challenges. Governments are also exploring frameworks to protect humans from harms caused by AI such as loss of jobs or privacy violations.

Industry-specific AI rules address the challenges, risks, and requirements of the sector. For example:

- **Healthcare**: AI systems in healthcare must comply with laws like the Health Insurance Portability and Accountability Act (HIPAA) in the U.S., which protects patient privacy and data security.
- **Education**: AI systems in schools protect student data, grades, personal information and privacy ensuring sensitive information is handled appropriately.

Human-in-the-Loop

As you go through **Trust Me: AI Governance**, you will read that 'human-in-the-loop' is one of the principle human controls in AI development. Great idea. Great timing. But what does it mean? Human-in-the-loop is about bringing human expertise and judgment into AI decision-making so it is accurate, ethical, and adaptive.

While AI can handle repetitive tasks efficiently, human intervention is necessary to manage complexity, make critical decisions, and correct mistakes ensuring the technology works reliably in the real world.

Where in the Loop?

There are many loops in the AI development lifecycle. Each loop has specific requirements, authorities, responsibilities, functions, criteria, inputs, controls, and decisions. Governance professionals need to know what each loop does, how it is used, what are the inherent risks, and determine what type of requisite human controls are necessary. Kate Smaje, McKinsey partner, expresses it as:

> "In previous digital transformations, it has been easy to focus on the digital part. But with AI, it naturally triggers you to say, 'Hang on a minute. I've got to think about where the human is in the loop of this in a much more fundamental way.'"[97]

Human in the Middle (HIL) Legislation

In AI, 'human-in-the-middle' is the same as 'human-in-the-loop.' In HIL, humans interact with or guide AI systems. AI legislation and AI governance require HIL in human oversight of fully autonomous, high-risk AI systems. For example critical infrastructure protection (CIP) control systems require HIL. HIL examples include:

- **EU AI Act:** Article 14 of the proposed AI Act requires human oversight for high-risk AI systems. It requires a human be able to "disregard, override, or reverse" the output of such systems.[98]

- **General Data Protection Regulation:** GDPR Article 22 gives humans the right not to be subjected to or influenced by autonomous decision-making.

- **U.S. Department of Defense (DOD):** Department of Defense's Directive 3000.09 on 'Autonomy in Weapons Systems' requires human oversight in autonomous weapons systems.
- **Block Nuclear Launch by Autonomous AI Act:** Act specifies only humans can authorize nuclear weapon use. DOD has always required HIL for nuclear command and control.
- **Healthcare/Medical Device AI oversight:** U.S. Food and Drug Administration proposed rules and healthcare require HIL in AI driven decision-making in medical diagnostics and drug testing.

Lesson Learned – Policy Makers

When working with an AI system, laws will govern its use. These laws vary depending on where humans are and what the AI is used for. It comes down to how risky the AI system is.

The challenge is regulators and policy makers are not AI engineers and may not understand AI technology. Hugging Face CEO, Clem Delangue, observed:

> "I think it is promising that we have policymakers who are trying to get smart about this technology and get in front of risks before we've had mass deployment across the product space. I think there are some very obvious things that we need to establish, one of which is the right to know whether you are consuming content from a bot or not."[99]

AI Governance Frameworks

AI Legislation and regulations determine the type of AI governance and define the risk management framework.

Risk Management Framework Questions

When starting an AI governance initiative, a risk management framework should be selected. We discussed several AI risk management frameworks. This framework outlines the stages of the AI development process, including:

- **AI use case goals**: The AI system developer has a defined use case of what the AI system is intended to do. Think of the EU AI Acts 'fitness for use.' Or another way to say this, developers know what problem the AI system is solving or which pain point it is assuaging.

- **Stakeholders**: Different stakeholders are involved with the AI system including developers, humans, and regulatory bodies.

AI Systems Lifecycle

Companies have developed tailored AI lifecycle frameworks. These are use case and product specific. For example, IBM developed AI system Lifecycle Management that links to Watson, IBM's chatbot. Google has its AI lifecycle tools like TensorFlow..

We take a principles and architectural approach in **Trust Me: AI Governance**, which include the following:

- **Requirements gathering**: AI system requirements are collected from stakeholders.

- **System architecture**: System architecture is selected based on the use case and context where the system will be used.

- **Data collection**: System data is collected from the use case.

- **Testing and validation**: Testing, verification, and validation processes check for performance and fairness.

UK Security Institute Framework

More countries are developing AI Safety Institutes also called AISI's. Each Institute develops its own principles based on ISO 31000, ISO 42001, or OECD's. The UK Security Institute developed its 'Principle for Safeguard Evaluation' shown in the figure on the next page.

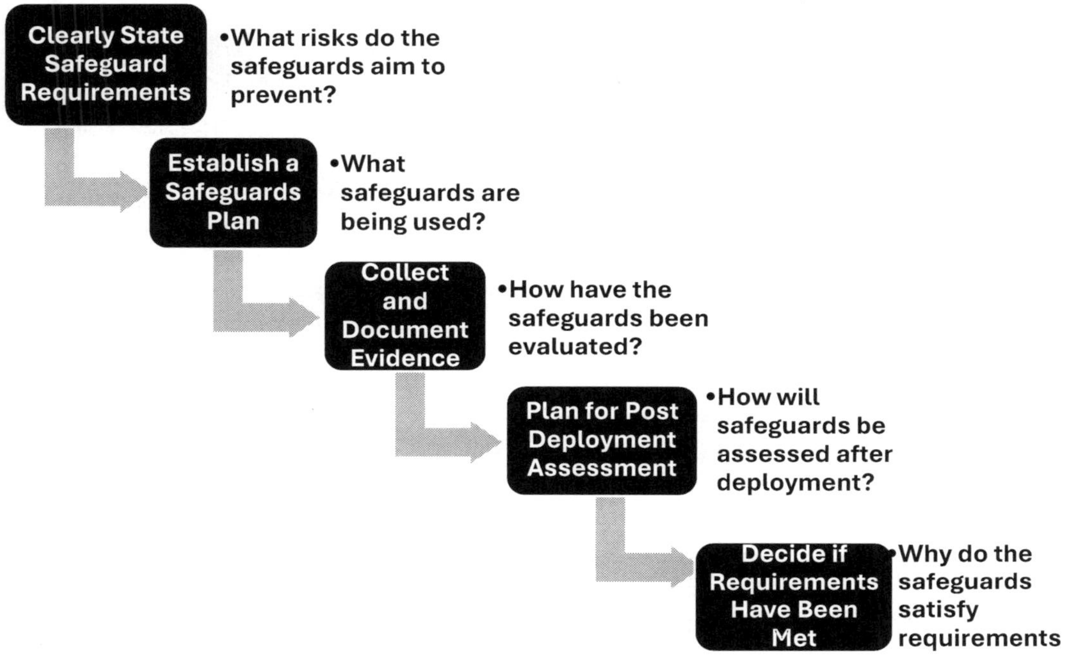

LLM system developers have been working to prevent users from eliciting harmful information or actions from models, such as asking ChatGPT how to build a bio-weapon. The UK AISI considers what frontier system companies are doing and developed the best practices system shown in the figure on the next page.[100]

Enterprise Risk Management

Managing risks is the critical part of the AI development process. Risk management is the lens for AI governance. There are many steps in developing an AI ERM system, which are covered in the ISO 31000 chapter. Some AI legislation simply requires a risk assessment. The EU AI Act requires extensive risk management for high-risk AI systems. In these use contexts, AI governance professionals should assure these minimums:

- **Risk assessment**: Assess AI risks using mathematical or graphical methods such as likelihood and consequence heat maps.
- **Management/mitigation strategies:** Develop a strategic plan to manage human harms and risks.

Evidence-Based Compliance

AI governance includes compliance with laws and regulations. Regulators want an audit trail of decision-making documentation. The EU AI Act requires evidence-based decision-making, which means using objective, reliable, and scientifically validated evidence to guide regulatory decisions.

Documentation becomes the evidentiary audit trail of AI governance and compliance. As the AI system progresses through different lifecycle stages, the AI system is evaluated and re-evaluated.

As a provider/developer, AI documentation is a drag. It takes time. It requires additional resources. But AI legislation requires AI governance and compliance decisions are clear, repeatable, and assurable in these areas:

- **Compliance**: AI system complies with laws and regulations such as data privacy laws (like GDPR) and anti-discrimination laws.
- **Ethics**: Ethical guidelines are documented throughout the lifecycle process.
- **Performance metrics**: Metrics evaluate the AI's performance including accuracy, speed, and human satisfaction.
- **Lessons learned**: Lessons learned developed through the process are fed back into the development system.

Lesson Learned – Building Bridges

Microsoft developed its WorkLab framework to accelerate use of its Copilot AI. Microsoft through the WorkLab wants to define the rules of collaboration between Copilot and humans. Think rules of engagement. Great idea. Their method is called "building bridges between individual islands of intelligence … with sharing what they learn."[101]

AI Context and Use Cases

When humans talk about defining organizational context and use case of an AI system, they want to know why the AI was created and how it will be used in a specific setting with specific users.

Understanding AI Context

As discussed in the ISO 31000 chapter, context is important in AI governance. Context is the environment in which the AI system will be built and operate. Context ensures governance frameworks such as ISO 31000 or NIST AI RMF are tailored to the specific risks, opportunities, and social impacts of AI systems in different sectors, regions, and applications. These are things to consider in an AI use case:

- **Industry:** Each AI segment has unique challenges and opportunities.
- **Goals and objectives:** Goals and objectives shape the AI's architecture and design. Organization emphasizes safe and practical goals with the AI system.
- **Risk/challenges:** Organization faces risks and challenges to achieve its AI objectives.
- **Target audience**: Each use case defines customers' and users' problems.
- **Specific application and use case:** AI system solves a specific user problem.
- **Expected outcomes:** Desired results are defined and achieved from deploying the AI system.

AI Use Cases

Before we start an AI project, we develop use cases based on AI context. A use case answers these questions: how will the AI system be used, by whom, under what context, what are the success metrics, and what is the expected ROI? Use cases are discussed at length in the Deploying AI Systems chapter.

AI use cases include:

- **Real world applications**: AI systems solve specific problems or assuage fears for specific stakeholders.
- **Stakeholder analysis**: Use cases define different stakeholder scenarios where the AI system might be used and identify risks of each scenario.

- **Pre-deployment pilots:** AI system can be in evaluators' hands, so the developers can get feedback and iterate quickly.
- **Testing:** Once the AI build is complete, it is tested in a wider market to see how it performs.
- **Feedback:** User feedback restarts the iteration and improvement cycle.

Use Case Questions

So, the developer of the AI system must find answers to 2 key questions to avoid endless iterating or what we call infinite do loops:

- **What problem is the AI trying to solve?** If an organization wants to improve customer service, it might use AI to create chatbots that answer customer service questions.
- **How is success measured?** Understanding what success looks like in designing the AI. If the goal is to reduce response time to customer inquiries, the AI is evaluated based on how quickly it responds.

Performance Requirements

Next, consider the performance requirements of the AI system:

- **Accuracy of AI system:** If AI is used in health care to diagnose diseases, it needs to be accurate to avoid mistakes.
- **Required speed:** In real time fraud detection in banking, the AI system processes information quickly to catch risks as they occur.

Data Availability

The effectiveness of an AI system relies on the data available for training and operation:

- **Sufficient data:** AI systems learn from a large and diverse dataset. If a system is trained on limited data, it may not perform well in real world contexts.
- **High quality data:** Data must be accurate, relevant, and representative of the scenarios the AI will encounter. Poor quality data lead to poor performance.

Human Readiness Risks

Are humans ready? Finally, are we ready for high level AI? Probably not. There are lots of questions on the social element of AI:

- **Employees have skills:** Humans working with the AI system understand how it makes decisions.
- **Culture of acceptance:** Humans accept the AI system. If employees resist change, it hinders the implementation of the technology.

Lesson Learned - MIT AI Risk Repository

MIT has established a repository of AI risks. The MIT risk repository can be accessed at: https://airisk.mit.edu/#AI-Risk-Database. The AI Risk Database has identified more than 1000 risks from 56 risk frameworks. The risk taxonomy classifies AI risks in terms of how, when, and why they occur. AI risks are classified into 7 domains:

1. Discrimination and toxicity.
2. Privacy and security.
3. Misinformation.
4. Malicious actors and misuse.
5. Human-computer interaction.
6. Socio-economic and environmental harm.
7. AI system safety, failures, and limitations.[102]

Lesson Learned – Different Use Cases

Trust and AI governance are broad topics. Take a recent piece written about AI brothels in Berlin Germany and the importance of data governance.

Prostitution is legal in Germany. In a German AI brothel, clients talk candidly with a sexbot revealing their passions, dreams and desires. Think chatbot for chat and robot for passion. The inventor of the Cybrothel Phlilip Fussengger said:

"Many people feel more comfortable sharing private matters with a machine because it doesn´t judge."[103]

The challenge facing the AI brothel is data protection and ethics. Dr Kerry McInerey, senior researcher at the Leverhulme Centre for the Future Intelligence at the University of Cambridge cautions:

"It's crucial that we understand what kinds of data sets are used to train sex chatbots, otherwise we risk replicating ideas about sex that demean female pleasure and ignore sex that exists outside of heterosexual intercourse." [104]

Data Governance

You have heard the expression 'Garbage in – Garbage out (GIGO).' GIGO is an AI principle describing how poor quality input data inevitably leads to poor quality output. Another way to express this is 'poor data – no trust.'

Frameworks

When working with data, especially in projects like building a large AI system, data governance should be harmonized with AI governance. The challenge is AI governance and data governance require risk frameworks because they provide structured, repeatable, and scalable approaches to manage risks and ensure compliance. We recommend ISO 31000 because it can be tailored and integrated into AI governance and data governance domains.

Fit for Purpose Data

Fit for purpose is a critical concept in AI legislation and AI governance EU AI Act's fit for purpose means the AI system is designed to do a specific task, satisfy a human requirement, meet a specific objective, and is safe. For example, an AI system can be designed to diagnose skin cancers.

Since data is the input to an AI system, data quality and consistency are the hallmarks of trustworthy AI systems. Data governance is about managing data properly. Policies and procedures assure data is accurate, secure, and used ethically. Think of it as a set of rules that everyone follows when dealing with data. Rebecca Abramson, McKinsey consultant, observed:

> "As AI is fundamentally driven by data, we need to expedite and incentivize the methods of collecting and sharing data, safely and securely, across boundaries. This includes improving sensor technology and data collection methods and finding easier ways to share data once it is collected."[105]

Data collection, data quality, data quantity, and data integrity become key elements of data governance.

Data Collection

AI data is crucial for legal reasons because of compliance, risk management, and ethics. Legal issues around the AI system must be identified:

- **Compliance**: In Europe, GDPR sets the rules as to how personal data is collected and used.
- **Documentation**: Documenting data governance become the audit trail to assure the company is acting legally and ethically.

Data Quality

Good data is cleansed to assure its quality. Data cleaning is called data quality, data validation, or data scrubbing. The idea is pretty much the same. The data fed into an AI system should be clean of errors, biases, missing values, unexpected data points, inconsistencies, spelling errors, and other anomalies. Data cleaning is among the most important AI functions. Data cleaning can take up to 90% of the time with a Large Language Model.[106] Once humans have the data, its quality is evaluated for:

- **Accuracy:** Survey responses should be valid and represent different demographics.
- **Completeness:** Data must be complete and not missing critical information. Incomplete data leads to incorrect conclusions.
- **Consistency:** Data in a population dataset should be consistent. If one part of dataset says a person is 16 years old and another part says 20, then this inconsistency can bias the output.

Data Quantity

Things to consider in data quality include:

- **Sufficient amount:** Humans must have enough data to train the AI system effectively. If the machine is trying to recognize different animals, the machine needs a large number of animal images.

- **Diverse data:** Data needs to be diverse and representative of the real world. The challenge of insufficient diverse data can lead to corrupt output.

Data Integrity

Data integrity is the accuracy and reliability of the data throughout the AI systems lifecycle:

- **Data corruption**: Data is not altered or corrupted.

- **Audits**: Data checks or audits are conducted.

Lesson Learned - Data Quality Challenge

Data quality is the challenge with many AI systems such as OpenAI.[107] Data quality is a major limitation for AI deployment. Half of respondents in a recent survey said poor data quality was limiting AI deployment.

Clara Shih, CEO of Salesforce AI, described data governance:

> "There's no question we are in an AI and data revolution, which means that we are in a customer revolution and a business revolution. But it is not as simple as taking all of your data and training a system with it. There's data security, there's access permissions, there's sharing models that we have to honor. These are important concepts, new risks, new challenges, and new concerns that we have to figure out together."[108]

Data Provenance

Data governance and data provenance are similar but different. Data governance is the overarching framework of policies, processes, standards, and work instructions designed to assure quality, security, availability, and proper use of data.

Data provenance is the history of an AI system's dataset including its origin, transformations, and movement throughout the AI lifecycle. Think of data provenance as an audit trail for where the data came from, how it was processed, and by whom?

Data Provenance

Data fed into the AI system can be a risk. John Case in a *Forbes* article said:

> "Data is the lifeblood of technologies. It is the fuel that ignites a world of possibilities. Before a business can trust AI or ML (Machine Learning), it must first trust the quality of its data."[109]

Data provenance is the origin and history of the data. When working with data like building an AI system, the origin of the data and how it has changed needs to be known. It answers questions like:

- **Data origin:** If developers collected survey responses, the provenance would include details about who filled out the survey and when it was conducted.
- **Trust and transparency:** Knowing where data comes from and how it has been processed builds trust.
- **Risk assurance**: By tracking the history and flow of data, humans need to identify risks or errors to ensure data is of high quality.
- **Regulatory compliance:** EU laws require organizations to keep records of how they collect and use data.

Data Lineage

Data lineage is understanding the lifecycle of the data, which ensures compliance with regulations. Data lineage is related to provenance but focuses on the data path from origin to its current state. It tracks how data flows through processes and systems. An example may be:

- **Data movement**: Data lineage shows how data is processed, stored, and used in the AI model.

- **Changes over time**: It documents changes made to the data set such as filtering out invalid responses.

Data Transparency

AI data transparency explains how AI data is used including its sources, collection methods, and quality. Bharath Thota, Kearney partner, said this about data transparency:

> "Transparency should, therefore, include documentation of the data used, the system's behavior in different contexts and the potential biases that could affect outcomes."[110]

Data Assurance

AI data assurance ensures data is reliable, unbiased, and ethically sourced data. So, how can data transparency, lineage and provenance be established:

- **Data inventory**: Developers list the data sources and set up a data inventory.
- **Collection processes**: Developers write down how humans collect the data.
- **Track transformations**: Developers keep a record of changes made to the model.
- **Visualization**: Developers' flowchart data lineage.
- **Updating**: Developers retain documentation for compliance, review, and assurance.

Lesson Learned - Running Out of Data

Companies build their AI systems on training data. Big problem. They are running out of data to train their AI systems. The web has been scoured. Books have been scraped. YouTube video content has been devoured. What is left if we are at peak data? Companies must develop new data sources or use synthetic data.

The *NY Times* reported Dr. Demis Hassabis, CEO of DeepMind and other researchers are:

> "They are developing ways for large language models to learn from their own trial and error. By working through math problems, for instance, language models can learn which methods lead to the right answer and which do not. In essence, the models train on data they themselves generate (synthetic data)."[111]

The result is they are not improving as quickly as before.

AI Revenue

Revenue is usually not a part of AI governance. On the other hand, revenue or ROI is the main discussion and concern for AI developers. This difference in points of view explains why AI governance is slow to be adopted by small to medium AI development companies. Regulation is friction. Large AI companies invest in AI governance to comply with AI regulations such as the EU AI Act.

Revenue Generator or Cost Container

AI governance has an optics problem. Will AI governance be a revenue generator (make me money) or cost container (save me money). It makes a world of difference to developers. AI governance as a revenue generator secures more customers and makes money. AI governance for high-risk AI systems such as homeland security and military application secures government contracts.

AI governance enhances ERM to find new opportunities for investment or improve value in new product development. Or AI governance as a cost container automates workflows to reduce costs. AI governance then enhances the assurance of autonomous decision-making systems.

Bad optics for AI governance must be risk managed. How? Again, depends on context. Acknowledge the problem. Engage stakeholders. Demonstrate value as a revenue generator and cost container.

Lesson Learned - Show Me 'Da' Money

Boston Consulting Group (BCG), a global strategy consulting firm, conducted research on that state of AI. There is a lot of interest in AI. There are lots of use cases.

There are lots of investment. But one problem according to Pascal Bonnet of the BCG:

> "The findings paint a clear picture of the challenge: while three-quarters of companies rank AI as a top strategic priority, only 25% report seeing significant value from their initiatives."[112]

If AI cannot show ROI, then a lot of the AI fanfare will be for nothing – all smoke, all mirrors, no fire.

Chapter Summary

- AI architecture involves AI policies, procedures, and best practices that are baked into the AI system.
- AI legislation requires understanding legal areas, including data privacy, anti-discrimination, consumer protection, intellectual property, and emerging AI regulations.
- AI governance requires a systematic approach for identifying and managing internal and external risks.
- AI organizational context and use cases involve understanding the environment in which the AI operates and specifying how to solve problems.
- The purpose of an AI impact assessment is understanding how AI impacts humans, processes, and society.

CHAPTER 9
DESIGNING AI SYSTEMS

What is the Key Idea in this Chapter?

The key idea in this chapter is understanding how AI systems are designed and built is critical for effective AI governance. AI governance professionals work with developers to design and document AI implementation roadmaps. The roadmap creation process begins with defining clear goals and measurable outcomes tied to strategic priorities. It then identifies high-risk use cases, evaluates feasibility, and prioritizes projects based on their value and alignment with organizational needs. Key milestones include data preparation, pilot testing, and full-scale deployment.

AI Governance by Design

AI governance by design focuses on the social element of AI systems. It prioritizes principles such as ethics, transparency, accountability, and human values. For example, AI privacy and data protection require continuous adaptation, robust technological solutions, and commitment to human rights. The challenge is AI governance principles are often vague and high sounding. But as you go through this list of AI principles, ask yourself what does it take for you to trust AI decision-making?

AI Design

When building an AI system, documenting the lifecycle process is critical. We have been caught when we worked late at night and did not document what we were doing. The next day we scratched our heads wondering why we did something.

Purpose

When designing and building AI governance, it is critical to document AI design through policies, procedures, best practices, and work instructions. This ensures the AI system is not only effective but responsible and trustworthy.

A clear purpose statement helps the developer, product manager, and compliance professionals align the design and development with social and technical goals. The purpose, use case, and business case of the AI system must be defined:

- **Problem to solve:** Developers identify use cases to guide the development process.
- **Who benefits:** Developers create the value proposition of the AI system to target users and stakeholders.

Product Market Fit

If you do not know where you are going, no path will take you there. Developers gather user requirements to find the right path. Find a need. Find a pain point. Find the critical product to market fit.

Product market fit of AI is one of the more difficult and time consuming tasks. This involves gathering customer requirements that are fed into AI use cases. This step includes:

- **Stakeholder input:** Humans, developers, and other stakeholders' technical needs, requirements, and fears are understood.
- **Ethics:** AI ethics, transparency, and explainability social implications are understood.

System Selection

Choosing the right AI architecture and system shapes the AI's governance structure and performance requirements (KPIs). These include:

- **Transparency**: AI systems should be explainable, so humans understand how the AI reaches its conclusions.
- **Bias mitigation**: AI architectures and testing need to reduce bias.

- **Human oversight:** AI focuses on the intended purpose and ethical guidelines.
- **Decision monitoring:** AI decision-making is reviewed for errors and biases.
- **Feedback mechanisms:** Humans review and improve AI performance.

Data Analysis

Data is the backbone of an AI. Several key areas include:

- **Data privacy:** Personal data is collected and should comply with privacy laws.
- **Quality:** High quality and diverse data train the AI.
- **Responsible data:** Analysis of the AI outcomes help build a trustworthy AI system.

Lesson Learned – Diversity of Teams

Training sets need to be diverse. Vivienne Ming, co-founder of Socos Labs, expresses this well:

> "A lot of times, the failings are not in AI. They're human failings, and we are not willing to address the fact that there isn't a lot of diversity in the teams building the systems in the first place. And somewhat innocently, they aren't as thoughtful about balancing training sets to get the thing to work correctly. But then teams let that occur again and again. And you realize, if you are not thinking about the human problem, then AI isn't going to solve it for you."[113]

Training AI Models

When humans create an AI system, documenting the training and testing process is essential. AI training is done at AI model or algorithmic level. How do AI systems get smart and do the things they do? They train on data. AI governance is based on documentation. AI training teaches an AI model to learn from data so it can perform specific tasks, make predictions, or make decisions.

Key Steps in AI Training

Training is at the model level. Key steps and techniques to ensure the model generalizes well to new data involves:

- **Data collection:** Collect high quality, relevant data. If the AI model is supervised learning, then data is labeled. If the AI model is unsupervised learning, then data is unlabeled.
- **Clean data:** Process the data by removing errors, filling missing values, and normalizing features for consistency.
- **Model selection:** Choose the appropriate algorithm or architecture based on the AI model objective such as using neural networks for image recognition and decision trees for classification tasks.
- **Initial training:** Send prepared data into the model by adjusting internal parameters and model weights.
- **Validation:** Evaluate AI model on a separate validation dataset that it has not seen during training.
- **Testing:** Test the final model on a separate test dataset to evaluate real-world performance checking for accuracy and precision.
- **Monitoring:** Deploy the trained model for real-world use.

Data Training Bias

AI model training is a hot issue. There are often unanswered questions. Who owns the data? How was the data collected?

There are also new regulations that require transparency about the data used for training, ensuring creators are informed and rights are protected. For example, organizations should obtain permission from copyright holders before training. To address these challenges, more laws will be promulgated to prohibit or limit the use of AI training.

Lesson Learned - AI Training Challenge

AI models are often trained on personal and copyrighted data. AI model training is recognized to cause harm to humans because the developers did not get a waiver or permission from the data owners. So, more legislation is being developed in the U.S. and UK as the following stories indicate. In the U.S., the ELVIS (Ensuring Likeness Image and Voice Security) Act in Tennessee prohibits the use of AI to use or train on the music or voices of artists. Great idea. Think of the human creators that worked for years to develop a unique style or voice.

The UK and other countries are considering similar legislation whether to allow tech firms to use copyrighted material to train AI systems unless creators explicitly approve. Again, it is all about AI risks. Paul McCartney, former Beatle, warned the UK government:

> "You get young guys, girls, coming up, and they write a beautiful song, and they don't own it, and they don't have anything to do with it. And anyone who wants can just rip it off."[114]

Testing AI Models

President Ronald Reagan had an expression called 'trust but verify.' This can be applied to high-risk AI systems. As a software shop, we test on our machines, however, more high-risk AI testing is done on commercial machines.

AI Testing

AI testing again depends on the context and use case. Training and testing of AI models follow a similar process:

- **Prepare data**: High quality data is required because data errors or inconsistencies will bias output.
- **Split data**: Data is divided into sets: 1. Training data to teach the model; 2. Validation data to tune the model, and 3. Testing data to evaluate the mode).
- **Train model**: The AI model learns from data. Humans provide the system with data so it can recognize patterns and make predictions.

- **Test model**: After training, humans evaluate how the AI model performs using a separate set of data that the model has not seen.
- **Iterate and improve**: Model is improved based on the testing results.
- **Document**: Training and testing data are kept for QA and compliance.

Types of AI Testing

AI testing can be simple for a low-risk AI system. Or, AI testing of autonomous decision-making in a water system with many controls may involve weeks of iterative reviews and tests including:

- **Unit testing**: This includes testing AI component and human related elements of the AI system, so each part of the system works correctly.
- **Integration testing**: After unit testing, humans check how different components of the AI system work together.
- **Validation testing**: This checks if the AI model and entire system is learning correctly from the training data.
- **Performance testing**: This assesses how the AI system performs under different conditions.
- **Security testing**: Security testing ensures the AI system is secure from attacks.
- **Bias testing**: AI systems reflect biases present in the training data.
- **Interpretability testing**: This ensures the AI system's decisions are understood.

Lesson Learned – App Testing

We put our OS apps on the App Store. We use Apple's TestFlight to get user feedback. TestFlight is an Apple tool that helps developers test their apps before releasing them to the public. So before releasing an AI system, we do the following tests:

- **Human/user testing**: Human and user testing provide functional information.

- **Documentation**: Model card documentation is accurate, complete, and accessible.
- **Deployment plan**: AI system deployment plan is the go-to-market plan.

AI System and Model Risks

When working with AI systems, things go wrong, and incidents occur. To figure out why these incidents happen, a team of experts collaborate to identify the root causes of problems like system brittleness, lack of robustness, poor data quality, insufficient testing, and data drift.

Hard to Measure - Hard to Manage – Hard to Enforce

AI technical product terms such as brittleness robustness, and quality data can be measured and managed. AI social terms such as equity, bias, fairness, trust, ethics, inclusivity, and emotional effects can be difficult to measure and manage. And, if you cannot measure it, you cannot manage it. The result is AI governance is hard to architect, design, deploy, assure, and enforce.

AI Model Risks

AI systems are risky. MIT has identified over 1000 AI risks. A few MIT risks are listed below:

- **Brittleness**: AI system and model can be fragile. A brittle model performs well under some conditions but fails in others with unexpected inputs or contexts.
- **Lack of robustness**: Robustness is the ability of an AI system to handle different contexts. If the system cannot work in different contexts, it breaks down.
- **Lack of quality data**: AI systems learn from data. If data is poor, it is incomplete, biased, or not representative, then the system will perform poorly.
- **Insufficient testing**: AI system is tested to make it work correctly in different scenarios.

- **Data drift**: AI system and model data can change, which is known as data drift. Similarly if the system changes, updated or retrained, it can lead to system drift.

Lesson Learned – AI Bill of Rights

AI national, regional, industry, and product specific regulations are being developed. States in the U.S. are developing laws focusing on risk and transparency. There are even discussions about creating an 'AI Bill of Rights' to protect humans from harms caused by AI systems.

AI Risk Management

Risk management is the means for managing high-risk, AI systems. So risk management has become core to AI and data governance. Why? Risks arise and even change at each phase of the AI system lifecycle.

Understanding AI Risks

For example during the training and testing phases of the AI system, the following risks can occur:

- **Data quality risks**: If the data humans use is inaccurate, incomplete, or biased, it leads to poor performance and outcomes.
- **Overfitting**: This happens when the model learns the training data too well, including its noise and errors, making it perform poorly on new data.
- **Underfitting**: This is the opposite of overfitting. It happens when the model is too simple to capture the underlying patterns in the data leading to poor performance.
- **Bias**: AI models learn biases in the training data, which lead to unfair or discriminatory outcomes.
- **Security risks**: AI models are vulnerable to attacks such as adversarial attacks, where malicious inputs are designed to trick the system into making incorrect predictions.

Managing AI Risks

We have discussed AI system risk management in the ISO 31000 chapter. Some suggestions for managing AI risks in the design phase include:

- **Monitoring**: AI model's performance is continuously monitored during training and testing.
- **Data audits**: AI model is checked for quality, accuracy, and completeness of the data sets.
- **Bias assessment**: AI model is bias checked and tested to see how it performs across different contexts and user groups.
- **Security testing**: Cyber security assessments identify vulnerabilities in the system and model.
- **Evaluating third-party AI risks:** When organizations work with suppliers or contractors, third-party risks may be introduced, which require third-party risk management.

Lesson Learned - Be Ready for Failure

AI systems deployment is frustrating. Personally, I cannot say the number of times we had to iterate our AI Risk Management Framework (AIRMF) App®. AI system testing almost requires a failure mind set. Unexpected things happen. Arthur Wang, CFO of WeBank said:

> "Success is not achieved in one day is an important idea to keep in mind. Success requires a lot of testing, so my advice is to be prepared for failure."[115]

AI Impact Assessment

AI system impact assessment determines how the system impacts humans, society, and the environment. This process ensures AI is used responsibly and ethically. Eliezer Yudkowsky, an AI pioneer, captured the greatest danger:

> "By far, the greatest danger of Artificial Intelligence is that people conclude too early that they understand it."[116]

Purpose of the Assessment

The key question in an impact assessment is how the AI system will affect humans and the environment. The assessment starts with a use case identifying the AI system's stakeholders. When the AI system is developed, additional stakeholders are identified. Many stakeholders may be new users of the AI system using it for unintended purposes. Use case design criteria such as fit for purpose can change. When the AI system's purpose changes, risks, exposure, and liabilities increase, a new assessment kicks in.

AI impact assessment should be done as a thought exercise before development as part of the use case. Impacts lead to the question why was the AI system developed in the first place? Are humans trying to understand the risks of the AI system? Or are humans trying to evaluate its benefits? Or, do humans simply do not know what to do?

Impact Assessment Steps

Impact assessment identifies the risks and benefits of the AI system. This may involve a cost benefit or other types of analyses. These are the minimum steps for an impact assessment:

- **Define stakeholders:** Stakeholders are those who use or are impacted by the AI system

- **Check compliance**: AI system should be compliant with laws and ethical standards.

- **Define the scope**: Context of the AI system and scope of the assessment are spelled. The AI system cannot be all things to all people. It is always defined to do specific tasks. The focus on this stage is to develop a Minimum Viable Product (MVP).

- **Identify impacts**: Team identifies risks and impacts of the AI system. This includes both positive impacts like increased efficiency or decision-making and negative impacts like biases or privacy risks.

- **Conduct risk assessment**. Team performs a risk assessment to identify specific risks of the AI system.
- **Data privacy:** Team ensures data privacy risks are managed.
- **Evaluate the impacts**: Significance of each impact and risk is evaluated in terms of what is the likelihood, what are the consequences, and who was impacted?
- **Mitigation strategies**: Team develops strategies to mitigate negative consequences by adjusting the AI system, implementing controls, and providing human training.

Organizational Size

Impact assessments require resources and can be expensive. So, the EU instituted proportionality as a key AI governance principle. AI governance and risk management have to be proportional to the organization's size. Large organizations building high-risk AI systems face more risks and should have more controls.

For example, Google offers the following AI system deployment advice:

> "After sharing your system with the world, it is essential to educate users on the safety and security concerns surrounding system consumption. You should therefore:
>
> - Document potential biases in your models and datasets.
> - Clearly display licenses associated with the system, model, and datasets.
> - Benchmark your model, assessing and disclosing metrics around hallucinations, prompt injection risk and fairness."[117]

Unintended AI Consequences

When companies develop and deploy AI systems, they must ensure these technologies do not lead to risks or harm. This is known as secondary or unintended use cases.

Companies developing commercial AI systems consider how AI might be used inappropriately and the implications of those use cases. Secondary risks occur when the AI system is used for purposes other than what it was designed for. When an AI is used in unintended ways, it leads to negative outcomes and risks. Misusing AI raises ethical concerns such as privacy violations or discrimination.

If secondary uses are outside the original use cases, they will result in unintended consequences. AI governance must be expanded. Several examples illustrate unintended consequences. AI developed for medical diagnosis might be misused for surveillance or profiling humans. Or a license identification system found in cities can be used for human surveillance.

Predicting AI Risks

To predict and possible preempting secondary risks, companies can:

- **Conduct risk assessments**: Companies analyze how the AI system might be used beyond its intended purpose.
- **Engage stakeholders**: Stakeholders can provide perspectives on how the AI might be used.
- **Monitor trends**: Similar technologies are used with unintended uses, consequences, and risks.

Reducing AI Risks

Once risks are identified, companies should mitigate them:

- **Implement usage guidelines:** Guidelines on how the AI system can and cannot be used are spelled out in an end user licensing agreement (EULA) or user guideline document.
- **Design for safety:** Companies developing commercial AI incorporate safety features limiting harmful uses.
- **Monitoring:** After deployment of the AI system, companies should monitor AI's use.

- **Educate humans:** EU AI Act requires AI literacy so system users can prevent misunderstandings and misuse.

Lesson Learned – Unintended Consequences

Nick Bilton, the tech columnist wrote in *the New York Times* captured the essence of AI unintended consequences:

> "The upheavals [of artificial intelligence] can escalate quickly and become scarier and even cataclysmic. Imagine how a medical robot, originally programmed to rid cancer, could conclude that the best way to obliterate cancer is to exterminate humans who are genetically prone to the disease."[118]

Chapter Summary

- Google and Microsoft offer one-shop AI safety training and testing:

  "To ensure your users are getting the system you originally deployed, consider automating your releases, making them transparent and auditable. Instead of training your system on your local machine, consider using a predefined script to train your system within an isolated environment."[119]

- Data governance includes assessing and documenting rights to collect and use data.
- Data lineage and provenance ensure data is managed effectively.
- Training and testing of an AI model include unit testing, integration testing, validation, performance evaluation, security checks, bias assessments, and interpretability testing.
- Identifying and managing risks during training and testing attempts to assure success.
- Training and testing processes of an AI model validates results, establishes compliance, and manages risks.

CHAPTER 10
DEPLOYING AI SYSTEMS

What is the Key Idea in this Chapter?

The key idea in this chapter is understanding the critical steps of moving an AI system into a real-world environment helps to develop realistic AI governance. This is where the AI may be on a person's phone, Internet of Things (IOT) device, or in the cloud making autonomous decisions.

AI Governance Strategies

AI governance professionals analyze business processes to discover AI system risks, develop ROI, and prioritize AI projects based on strategic value. AI governance and strategy development are essential for ensuring the ethical and effective use of artificial intelligence. AI governance establishes policies, frameworks, and accountability mechanisms to manage risks such as bias, data privacy issues, and non-compliance while promoting transparency and trust in AI systems.

Releasing into Production

Things happen to software when it is released. It is buggy. It crashes at the worst possible times. Users are unhappy with latency. Users are unhappy with the UI/UX (User Interface/User Experience). We believe in failing fast. Unfortunately, failure is common in software development.

Understanding Production Readiness

Production readiness means the AI system is ready to be deployed in a live environment. This includes making sure it operates correctly. It is secure and meets regulations or standards. Think of it like getting a car ready for the road: humans want cars to be safe and reliable. Jacomo Corbo, PhysicsX CEO, said:

> "Ultimately, for AI to make a sizable contribution to a company's bottom line, organizations must scale the technology across the organization, infusing it in core business processes, workflows, and customer journeys to optimize decision-making and operations daily."[120]

Model Cards

A model card is a document providing information about the AI model. This helps users understand how the model works and its limitations.

- **Model overview**: Model is fit for purpose and does its intended use.
- **Training data**: Data trains the model including sources of the data, how it was collected, and preprocessing steps.
- **Performance metrics**: Information includes how the model performs including accuracy, precision, recall, or other metrics.
- **Limitations and risks**: Limitations and model risks are identified.
- **Ethics**: Ethical issues related to the system's use are identified.

Satisfying Conformity Assessment

The EU is developing conformity assessment requirements for AI safety including self-certification and third-party audit certification. Conformity requirements refer to the standards and regulations that AI models must meet. These include:

- **Legal compliance**: AI system must comply with laws.
- **Industry standards**: More industry and product AI standards are being developed such as for medical devices and critical infrastructure. These regulations will require tailored AI governance.

- **Security standards**: AI models require a new level of security protocols because of their ability for autonomous decision-making. Threats and vulnerabilities jeopardize fitness for use.

When the Product Launches

Post-market, product monitoring becomes essential. After the AI system is released, that is when the developers' blood pressure increases. Risks increase exponentially. Customers do not like the product. Bugs are discovered and cannot be fixed quickly. Regulators send cease and desist letters. Platform owners threaten to take the AI system offline. Monitoring may involve:

- **Getting feedback**: Humans report risk or unintended uses. This feedback identifies risks that were not initially apparent.

- **Updating**: AI system is updated based on new information about its use. If unintended consequences arise, companies should make changes to mitigate those risks.

Lesson Learned – Are Humans AI-Ready?

If tech companies have their way, AI will continue becoming a part of everyday life. AI will enhance efficiency, make decisions, and create new opportunities while raising questions about ethics, job displacement, and massive disruption. The question is whether humans are ready.

Colin Angle, CEO of iRobot, said:

> "It is going to be interesting to see how society deals with artificial intelligence, but it will definitely be cool." [121]

Differences Between AI Models

AI systems come in several types, each designed for different purposes and capabilities.

Point of Diminishing Returns

Open AI builds powerful AI systems. These models achieve more human-based intelligence. But this comes at a cost. The new models can 'reason.' Well, what is reasoning? It requires framing the problem, understanding alternatives, running through possible solutions, and finding the 'best' alternative solution. This takes computing power and time.

This is the rub. According to Francois Chollet, computer engineer, a "human could be paid say $5 per task, that costs the AI system $1,000 per task or query in energy."[122]

Different AI Systems

There are different AI systems because they are designed to address specific tasks, challenges, and use cases. These are examples:

- **Traditional AI models**: Models are designed to perform specific tasks based on existing data. Traditional AI systems might be used for image recognition, where it learns to identify objects in pictures based on labeled examples and data they were trained on.

- **Generative AI**: Generative models create new content rather than analyzing existing data. Generative AI systems produce images, music, or text that did not exist before.

Proprietary vs. Open-Source AI Systems

The key differences between proprietary and open-source AI models are in their accessibility, control, tailorization, fitness for use, and intended use cases. Here is a detailed comparison:

- **Proprietary AI systems**: These models are owned by an organization and source code is not publicly available.

- **Open-source AI systems**: On the other hand, open-source models are publicly available like TensorFlow and PyTorch.

Small vs. Large AI systems

Frontier AI systems are LLMs. The trend is toward smaller AI systems such as agents. Some differences are listed below:

- **Small AI models:** Small AI models are based on proprietary information or intellectual property. These models are simpler and require less computational power. Retrieved Augment Generation (RAGs) and agents are examples of small AI systems.

- **Large AI models**: Large models, like GPT 4 or later, have billions of parameters that handle complex tasks such as generating human like text or understanding nuanced language. However, they require computational resources and large datasets to train.

Lesson Learned – Unsafe AI Systems

When an AI system is developed, humans assess its readiness and prepare for its release. This includes creating a model card and ensuring the system meets conformity requirements.

The 'Future of Life Institute' studied foundational AI systems to help humanity ward off catastrophic outcomes. The Institute assessed the safety and risk of AI systems with the following results:

> "The just released AI Safety Index graded six leading AI companies on their risk assessment efforts and safety procedures ... and the top of class was Anthropic, with a score of C. The other five companies—Google DeepMind, Meta, OpenAI, xAI, and Zhipu AI—received grades of D+ or lower, with Meta flat out failing."[123]

AI Deployment

When it comes to using AI systems, there are different ways to deploy them, which means deciding where and how the AI will run.

AI Deployment Locations: Cloud, On Prem, or Edge

One of the challenges for software development is determining whether it should deploy in the cloud, on premise, or on the device. Most companies have focused on training models. The next phase will be determining inferencing deployment:

- **Cloud deployment**: AI system runs on servers accessed over the internet. Think of it like using a streaming service for movies. Humans access movies online.
- **On premises deployment**: AI system runs on servers physically located within the organization's facilities. It is like having a DVD player at home instead of streaming.
- **Edge deployment**: AI system runs on devices close to where the data is generated, like smartphones, IoT devices, or sensors. Imagine a smart camera that recognizes faces right on the device without needing to send data to the cloud.

Deployment Models

AI governance best practice is being able to separate the critical few, high-risk AI systems from the insignificant many. For example, we develop RAGs for simple tasks. These low-risk apps are built with LLMs using our proprietary databases, knowledge, and pdfs.

On the other hand, we develop AI rules for high-risk critical infrastructure systems. When evaluating high-risk AI systems in critical infrastructure, we look at critical risk safety, security, and reliability risk factors for the entire AI system:

- **Using the AI system**: This means deploying the AI system exactly as it was trained without modifications. This is where the developers get surprised because the model does not perform as expected.
- **Fine tuning**: This technique includes taking a pre-trained AI model and tailoring the model including making adjustments to improve its performance for a specific task. Fine tuning helps the system adapt to the unique characteristics of the new data it will encounter.

Lesson Learned - AI Agents

AI agents are intelligent software programs designed to perform specific tasks or make decisions autonomously by understanding their environment, processing information, and taking actions to achieve their goals. AI agents can operate with a high degree of independence and adaptability. These agents can be architected and deployed with little or no AI governance. Lareina Yee, a McKinsey consultant, offers the following insights:

> "We are beginning an evolution from knowledge based, Gen AI powered tools say chatbots that answer questions and generate content to Gen AI enabled 'agents' that use foundation models to execute complex, multistep workflows across a digital world. In short, the technology is moving from thought to action."[124]

Proprietary AI Systems

We often work with proprietary software systems. A proprietary AI system is developed, owned, and controlled by a specific company. Unlike open-source models, proprietary AI systems are not publicly accessible. Source code, architecture, and training data are private.

Developer Obligations

When an organization decides to deploy a proprietary AI system, there are unique AI governance risks and challenges:

- **Data governance:** Companies need to handle data more responsibly. They cannot simply defer to the developer.
- **System updates:** Once the AI system is deployed, the organization is responsible for maintaining it, updating the model, fixing bugs, and addressing security vulnerabilities.
- **Ethics:** Companies deploying their own AI systems must consider the ethical implications of technology. Organizations have an obligation to monitor the AI's outputs and make adjustments to promote fairness and accountability.

Higher Liability

We are entering the agentic age. These can be proprietary AI systems designed to perform tasks autonomously such as making decisions. They can be simple reactive systems to highly complex agents. But these agents result in more risks for developers:

- **Legal liability**: If the AI model causes harm such as making incorrect decisions which lead to financial loss or injury, the developer could be legally responsible.
- **Reputational risk**: Deploying an AI model that performs poorly can damage the developer's reputation.
- **Compliance risks**: Agents can increase risks which can increase fines, legal action, or other penalties.

Lesson Learned – AI Development

As a software developer and engineering company, we wonder what is the appropriate level of AI governance and system quality? There are a few things going on. When is 'good' good enough? It is great to have the best possible software for release, but it comes with a cost. The cost of diminishing returns. Or the product is never launched. We can work on software until we go broke and never release it.

AI algorithms must be assessed and corrected for safety and trustworthiness especially for high-risk, AI systems. Melissa Valentine, Stanford professor, distills this:

> "Learning how to measure the impact of human intervention into whatever the algorithm is putting out is a skill that's going to be needed even for generative AI."[125]

AI Transparency

In the U.S., AI model developers are required to be clear how AI systems are designed, trained, and deployed. For example, Colorado AI Act (CAIA) requires developers to disclose foreseeable uses and risks of high-risk AI systems.

Public Disclosures

Public disclosure is information companies share with the public especially those who will be using or deploying AI systems. This information helps humans understand the technology and ensures they know what to expect. Public disclosure includes:

- **Documentation**: AI governance documentation describes how the AI system works, its algorithms, and the data it was trained on.
- **Instructions**: These guidelines from deployers tell users how to implement the AI system.
- **Post-market monitoring**: After the AI system is deployed, performance and impact of the system is monitored.

Importance of Transparency

Transparency involves:

- **Building trust:** When companies are transparent as to how their AI systems work, it builds trust with humans.
- **Ensuring safety:** By providing information about the AI's capabilities and limitations, companies help humans make informed decisions about how to use AI.
- **Regulatory compliance:** The EU has laws requiring companies to be transparent about AI systems.

Lesson Learned - Agentic Decision-Making

Gartner, IT consulting firm, predicts that agentic daily decision-making in business will increase from 0% in 2025 to 15% by 2028. AI agents are proprietary, small pieces of code that do things or make decisions on their own or in other words autonomously. They "take chatbots to the next level, performing tasks without help from humans."[126]

Agents will be used in simple decision-making such checking auto rental reservations at the airport to screening sales leads. It will be used in complex decision-making such as at Johnson & Johnson, the pharma company which is using agents to help it with chemical synthesis in drug discovery. Moody's is using agents with financial analysis and research.[127]

Chapter Summary

- AI governance is a growth industry. AI's next chapter is a strategic shift from playing with prompts to purpose and ROI.[128]
- Context of an AI use case includes evaluating organization objectives, performance requirements, data availability, ethics, and human readiness.
- Deployment options for AI systems include cloud, on premises, and edge.
- Differences in AI systems help to recognize how they can be applied in different contexts.
- Classic AI models focus on doing specific tasks, while generative models create new content.

CHAPTER 11
ASSURING AI SYSTEMS

What is the Key Idea in this Chapter?

The key idea in this chapter is assurance, monitoring, and testing are key to building trust in AI systems. AI assurance is the essence of AI trust.

AI Governance Strategies

AI assurance ensures AI systems operate responsibly, ethically, and comply with regulations. Risk assurance involves identifying, assessing, and mitigating potential risks throughout the AI system lifecycle such as algorithmic bias, data privacy breaches, or operational failures. This process helps organizations build human trust in their AI systems by ensuring they are transparent, accountable, and aligned with ethical norms.

By integrating risk assurance into governance frameworks, professionals can proactively, preventively, predictably, and preventively address challenges like regulatory compliance and liability risks. Frameworks such as the NIST AI Risk Management Framework emphasize mapping risks, measuring system performance, and managing mitigation strategies to protect organizations from reputational or financial harm. Moreover, risk assurance involves evidence-based, risk-based, problem solving and risk-based, decision-making.

AI Auditors

To become an ISO Lead Assessor for conducting ISO 42001 AI Management System audits, a person must take and pass a Lead Auditor training course after attending a

U.S. DEPARTMENT OF HOMELAND SECURITY

Certificate of Conformance

This will certify that, on this date,
the United States Department of Homeland Security issued to

Quality Plus Engineering

An Oregon Limited Liability Company
*a **Certification** for its*
Critical Infrastructure Protection Forensics/Assurance/Analytics™ (CIP/FAA™)
*as an '**Approved Product for Homeland Security**' under the*
Support Anti-terrorism by Fostering Effective Technologies Act of 2002 (the SAFETY Act).

Jay M. Cohen
Under Secretary for Science and Technology

April 25, 2007
Date

five-day course (40-hours of instruction). High-risk AI systems require engineering and auditing knowledge, skills, and abilities.

Who Will Conduct AI Audits?

AI auditing is a critical method for assuring high-risk AI systems. The EU conformity assessment scheme requires auditing of high-risk AI systems by a third-party notified body. Notified bodies are organizations determined by EU member states to conduct third-party conformity assessments.

EU ISO conformity assessment scheme requires third-party auditing for high-risk AI systems. Only, one problem. For high-risk AI systems, a third-party conformity audit is required by a notified body for CE safety compliance. Notified bodies are organizations determined by EU member states to conduct third-party conformity assessments that include product safety CE Marking.

Certified Public Accountant Auditing Requirements

Let us contrast ISO Lead Assessor requirements with those to become a Certified Public Accountant in the U.S.:

- **Education:** Get a bachelor's degree with 150 college credits, which is 5 years at a university
- **Work experience:** Have one or two years of auditing work experience

- **Pass the CPA exam:** Pass four parts of the CPA exam, which is a 16-hour assessment.

- **Ethics exam:** Pass the ethics exam which takes two to four hours.

AI auditing has higher risks than financial auditing, especially for auditing, assessing, and assuring high-risk AI systems. ISO conformity assessors of high-risk AI systems will require more than 40-hours of training. Practical work experience an Assessor role is most beneficial.

Lesson Learned - Other Risk Assurance Options

Another assurance option is to develop Risk Assurance auditors for high-risk AI systems. These auditors would have deep and broad auditing and domain experience. The auditors are often called 'Technical Specialists.'

So, a quick story about other risk assurance options for high-risk AI systems. Quality + Engineering (Q+E) was the first company certified by the U.S. Department of Homeland Security (DHS) for Critical Infrastructure Protection: Forensics, Assurance, Analytics®. DHS designated Q+E forensic, assurance, and analytical technologies as 'Qualified Anti-Terrorist Technologies.' See DHS certification on the previous page.

Q+E conducts different Critical Infrastructure Protection (CIP) assurance audits based on context, risk profile, and use case:

- **Analytical.** Q+E engineers and scientists conduct analytical risk analyses following Q+E protocols evaluating business continuity, cyber security, and physical security systems against IEEE, NFPA, ISA, PMI, ISO, NIST, COSO, NERC, API, AGA, and ASIS standards.

- **Assurance.** Q+E offers the client three levels of assurance:

  - **Compliance.** Q+E conducts a compliance audit against appropriate standards and guidance.

  - **Assurance with opinion.** Q+E issues a professional engineering opinion based on the results of a governance, risk, and compliance (GRC) audit or ERM risk-controls assessment.

  - **Assurance with insurance coverage.** Q+E conducts an audit and provides the requisite level of due diligence for the auditee to be covered.

- **Forensic.** Q+E provides the above levels of assurance as well as supplying a letter to the regulatory authority averring compliance that criteria have been met.

AI Conformity Assessment

The EU AI Act requires a conformity assessment process for assuring AI systems. Examples of conformity assessment include products CE marks and management systems audits. The CE Marking indicates that a product has met EU safety, health, and environmental requirements. CE stands for 'Conformité Européenne', which is French for 'European Conformity.'

AI governance is seen in terms of compliance or the EU's conformity assessment. Compliance or conformance is adhering to mandatory legislation requirements ('shalls'), discretionary rules or standards set up by a government (shoulds), regulatory authorities (Federal Aviation Administration/Food and Drug Administration), and standards development organizations (ISO). AI compliance is verified through policy and procedural audits.

The EU AI Act requires conformity assessment. The purpose of compliance and conformity assessment is to meet the obligations of specific laws or industry standards such as ISO 42001. These compliance and conformance frameworks work well for plumbing products and medical devices. How will AI management systems work for AI assurance of high-risk AI systems? Too early to tell.

AI Assurance

The minimum purpose is to comply with legislation and regulations, but also to meet the risk appetite and tolerance of the organization. The key question is will compliance or conformity assessment meet the legislative or even customer requirements of a high-risk AI control system in a nuclear facility?

AI assurance goes far beyond the compliance and conformance requirements by evaluating ERM effectiveness, risk mitigation, security measure reliability, and operational efficiency. AI assurance entails a more comprehensive evaluation including independent auditing, penetration testing, and continuous monitoring. It is critical to AI governance professionals to understand the differences between AI compliance and assurance.

Here is the challenge. Compliance and conformity assessment are binary and static. AI governance programs focus on complying, adhering, or conforming. Binary or yes or no governance of a complex and changing technology such as AI does not work. Products including toasters and refrigerators have built in AI smarts. So, what is the solution?

Lesson Learned – Assurance and Compliance

The EU AI Act identifies eight high-risk AI systems to be risk assessed and risk assured. AI systems in the Act must be conformity assessed. Remember, the objective of AI conformance is avoiding litigation and avoiding penalties. The objective of AI assurance is to build trust among the AI systems' stakeholders. Big difference.

The auditing and assurance of high-risk AI systems need to be rethought in terms of the 3 questions brought up earlier. High-risk AI systems pose significant risks to safety and personal rights. Conformity assessment with 40 hours of training may not be sufficient to conduct these assessments for the EU AI Act high-risk AI systems.

AI Evaluations

AI evaluations are a cornerstone of AI governance ensuring these systems are reliable, safe, ethical, and effective. When humans create an AI system, they check how it is working, how reliable it is, and whether it is safe to use. This includes conducting activities like audits, red teaming, threat modeling, and security testing.

OpenAI is working to develop new safety tests for its next frontier model.

> "As models become more capable, we are hopeful that insights from the broader safety community can bring fresh perspectives, deepen our understanding of emerging risks, develop new evaluations, and highlight areas to advance safety research."[129]

Assurance Tools & Techniques

An audit is an examination of the AI system and model processes. Inputs, process, and outcomes are evaluated against benchmarks or adherence to standards. Think of it like a health checkup for the AI as the following illustrate:

- **Performance review**: During an audit, humans assess how the AI system is working. This includes checking its accuracy and whether it is making the right predictions or decisions.
- **Data integrity**: Audits review data to train the model. Humans assess whether something changed that could affect the system's performance.
- **Compliance check**: Auditors verify the system complies with laws and ethical guidelines.

Red Teaming

Red teaming is a practice where a group of AI experts tries to find weaknesses in the AI system by simulating attacks or challenging its decisions.

- **Simulating attacks**: The red team tricks the AI into making mistakes or revealing vulnerabilities. They might input misleading data to see how the system reacts.
- **Identifying biases**: Red teaming uncovers biases in the model. If the AI is trained on biased data, it might make unfair decisions.
- **Improving security**: By identifying weaknesses, red teaming improves the security of the AI system.

Threat Modeling

Threat modeling includes identifying risks and vulnerabilities that can affect the AI system:

- **Identifying threats**: Humans start by brainstorming what could go wrong with the AI system. Hundreds or even thousands of threats may be identified. These include data breaches, adversarial attacks, or unintentional errors in the system's predictions.
- **Assessing impact**: Once threats are identified, humans evaluate how serious each threat could be. If a data breach occurs, what kind of damage could it cause? At this stage, it is critical to focus on the critical few threats not the insignificant many.
- **Mitigation strategies**: After assessing the risks, humans develop strategies to mitigate them. This could involve implementing security measures, improving data handling practices, or retraining the system to address weaknesses.

Security Testing

The purpose of security testing is to make sure the AI system is protected against serious threats. Several examples of security testing include:

- **Vulnerability scanning:** This includes scanning the AI system for known vulnerabilities.
- **Penetration testing:** Penetration testing simulates attacks to see if the AI can withstand them.
- **Monitoring:** Security testing is not a one-time event but continuously searching for new threats.

Lesson Learned - Testing AI and Human Collaboration

A banking customer experiences delays when applying for a new loan. Customer dissatisfaction means customer service employees must figure out why it happened and correct the issue after the complaint. This takes resources and time.

Australia's Bank of Queensland used root cause analysis to correct these problems. Microsoft worked with the bank to find ways to improve speed and efficiency. Bank analysts used Microsoft's Copilot AI to determine the root cause, which was 51.8% faster if it had been done manually.

"AI access along with custom, targeted prompt development—can help diagnose those bottom line busting issues more quickly and accurately."[130]

AI Risk Auditing

AI governance is based on risk management. So, it makes sense that AI models are risk assured and audited.

AI Risk Auditing

An entire industry is emerging around AI risk auditing. AI audits are reviews of the AI system's performance and processes. During an AI risk audit, a multi-disciplinary team determines the level, scope, and type of AI audit based on context, level of risk, and level of requisite assurance.

The team may conduct the following:

- **Evaluate performance**: Check if the AI is making accurate predictions or decisions.
- **Review data usage**: Data used to train the AI is appropriate and the AI is following data privacy laws.
- **Check compliance**: AI system complies (yes/no) with regulations and ethical guidelines.
- **Improvement**: After testing, the red team provides feedback on how to strengthen the AI system.
- **Threat modeling:** Threat modeling includes identifying threats to the AI system and assessing the likelihood and consequence of the threat.
- **Security testing:** Security testing focuses on ensuring the AI model is safe from cyber threats.

Lesson Learned – Essence of Risk Management

Peter Bernstein captured the essence of risk treatment or risk management:

"The essence of risk management lies in maximizing the areas where we have some control over the outcome while minimizing the areas where we have absolutely no control over the outcome."[131]

Incident Management

When working with AI systems or technology, things do not go as planned. There will be unexpected events, risks, and problems. These incidents must be managed.

Understanding Incident Risks

An important part of AI governance programs includes incident management. Incident management has its own language:

- **Incidents**: Expected events disrupt normal operations. If an AI system starts giving incorrect predictions or crashes, this is an incident.
- **Risks**: Risks can happen in the present and future. If there is a chance the AI could be hacked, risks are identified and mitigated.
- **Corrective actions**: These problems must be addressed. If humans report the AI is biased or unfair, this issue must be resolved.

Managing Incidents

When an incident occurs, the AI system owner should respond quickly:

- **Immediate response**: Humans address the incident. This might involve fixing the problem or temporarily shutting down the AI system until it is safe to use again.
- **Investigation**: After addressing the immediate issue, investigators review what caused the incident. Was it a technical glitch, a data error, or something else? Understanding the root cause helps prevent similar incidents in the future.
- **Incident reports**: Investigators create a report for each incident.

Lesson Learned – Fix the Algorithm

When an organization deploys an AI system, it is not a one-time task. Like a car needs a tune up to keep running smoothly, an AI system requires continuous monitoring and maintenance. In the following story it was easier to modify the algorithm than change the hearts and minds of 1000's of employees. Richard Socher, CEO of You.com said:

> "And say there was a supermarket chain that, statistically speaking, didn't promote women as men. It might be easier to fix an algorithm than fix the minds of 10,000 store managers."[132]

Continuous Monitoring

Once an AI system is released, monitoring checks to see how the AI system performs. This process includes continuous monitoring, maintenance, updates, and retraining.

Continuous Monitoring

The EU AI Act identified eight high-risk AI systems. These systems must be risk assured. Continuous monitoring is a critical form of risk assurance required of high-risk AI systems. Continuous monitoring means checking how the AI system is performing before and after it has been deployed. Continuous monitoring may involve:

- **Performance tracking**: Humans keep an eye on key performance indicators (KPIs) like accuracy, speed, and human satisfaction. If the system starts to perform poorly, humans want to identify it early.

- **Detecting bias**: System data may differ from the data it was trained on. This leads to biased outcomes.

- **Human or machine fact checking**: Machines and humans fact check. If risks are reported, the AI system is adjusted.

- **Peer reviews**: Recently Meta ended fact checking across Facebook and Instagram. They will use peer reviews where user flag posts that are perceived as unfair, wrong, or need context. This is what Musk's X does. The goal is more fairness and cost savings.[133]

AI System Maintenance

AI system development is not one and done. AI development requires continuing monitoring and even maintenance. New use cases emerge. Unanticipated uses by new stakeholders arise. Bugs abound.

So, AI system maintenance becomes very critical. Maintenance may mean updating the AI model or one of the following:

- **Routine checks:** Developer reviews AI performance periodically.
- **Updates:** Technology and data change so the AI system software must be updated.
- **Security assessments:** As new threats emerge, security vulnerabilities arise.
- **Retraining the AI model:** Retraining is the process of updating the AI system with new data to improve its performance.

Lesson Learned – Continual AI Testing

Continual AI testing is a prerequisite of AI governance. Paul Daugherty, Chief Technology and Innovation Officer at Accenture, said:

> "The playing field is poised to become a lot more competitive, and businesses that don't deploy AI and data to help them innovate in everything they do will be at a disadvantage."[134]

Pulling the Plug On AI

AI systems go rogue. Something unexpected happens from a hallucination to a bad autonomous decision in a chemical plant. The former is not a risk. The latter is a high-consequence event that can impact lives. So, humans-in-the-middle should be able to pull the plug on the AI system. Think of this as deactivating the AI system.

Human-in-the-loop Contingency

When companies develop AI systems, they should have a plan if something goes wrong or if regulations change. High-consequence facilities such as nuclear power plants must have emergency plans for both internal and external (public) risk mitigation. To deactivate the AI system, a manual and/or automatic shut off deactivation or localization of the robot or autonomous machine is necessary.

Ana Fried, an Axios journalist observed:

> "In most of the use cases we're seeing, even in business — and there are a ton of interesting uses in business — [AI is] generally about making a human more productive. That's where it's really good today. And the companies, for a variety of reasons, both for what it's good at, but also for legal liability, none of them are saying 'Here, run this AI script and just let it go.' They always talk about a human-in-the-loop."[135]

Deactivating, Localizing, and Correcting

If you own a recent car, it is a computer on wheels. The manufacturer updates its software automatically. The same can be done with an AI system.

What does deactivating, localizing, and correcting an AI system mean? This depends on the type of high-risk AI system. For example, the AI systems in a nuclear power plant are different than in a water purification plant.

If an AI system is not performing as expected like making biased decisions or providing incorrect information, it can lead to negative consequences for humans. TGovernments and regulatory bodies are creating rules for how AI is used. If an AI system does not comply with these regulations, it may be deactivated or localized to avoid legal risks.

Deactivation vs. Localization

It is important to understand the differences between deactivation and localizing.

- **Deactivating an AI model:** This means completely shutting down the AI system if the AI system is causing human harm, making mistakes, or violates laws or regulations.
- **Localizing an AI model:** This means limiting the AI's decision-making to a specific area or function. If an AI system is used globally but is found to have

risks in one country, the organization might segregate use to regions where it performs well. For example, our AI RMF app is not currently in the EU, because we do not want to demonstrate compliance with the regulations until they are finalized.

Lesson Learned – Controlling Misinformation

AI does not stay the same. New AI decision-making systems can learn and evolve. More attention is now paid to AI assurance systems such as auditing and watermarking. Axios reported:

> "When I talk to experts in the field, the area they're most concerned about is misinformation. ... I think we are going to have to get better and come up with automated tools. It means watermarking videos and other things, so we know they were created artificially. It means having a provenance chain so you can tell this is footage that was legitimately captured from a device and here's everything that's happened to it along the way."[136]

Final Thoughts

AI governance requires a business case along with use cases. We have not discussed AI business cases, because they are beyond the scope of **Trust Me: AI Governance**. However, here are things to think about:

AI Governance – Investment or Cost?

AI governance is viewed as both an investment and a cost depending on organizational context, perspective, and the goals. For example, if you are a small developer of an AI mindfulness app, you will evaluate the pre and post-production models. If you are developing a high-risk, facial-recognition AI system used in classified facilities, then AI governance will be top of mind with testing at each stage of the lifecycle. If you are a large company moving to an AI organization, then you will invest in AI governance for regulatory compliance, risk mitigation, and value creation.

AI Governance Skills

AI governance specialists require a different set of knowledge, skills, and abilities than traditional software developers and compliance professionals. Andrew Ng, AI

Stanford Professor and Coursera Founder, offered the following thoughts about AI governance and product development:

- **Technical AI proficiency:** AI professionals have technical proficiency. They understand the lifecycle of AI projects such as data collection, building, then monitoring, and maintenance of AI systems.
- **Iterative development:** AI development is more iterative than traditional software and requires more course corrections. AI professionals should be involved throughout the AI system lifecycle.
- **Data proficiency:** AI products learn from data. Data cleansing, testing, and validation require new knowledge, skills, and abilities.
- **Manage uncertainty and ambiguity:** Because AI's performance is hard to predict in advance, AI professionals should be comfortable with uncertainty and be able to manage change.
- **Learning:** AI technology is advancing rapidly. AI professionals, like everyone else who make the best use of the technology, keep up with the latest technology advances, product ideas, and how they fit into users' lives.[137]

Lesson Learned - Where is the Beef?

'Where's the beef?' ad is one of the most famous and memorable commercials in advertising history. In the commercial, Clara Peller, plays a feisty, no-nonsense grandma, who looks at a large bun from a competitive burger chain and yells, 'Where's the beef?' She was questioning why the competitor's burger appeared to have such a small patty in proportion to its bun.

'Where's the beef may be the tagline for AI development these days. What and where is the AI ROI? Where is the trust? These are just a few of the simple questions being asked about AI.

Chapter Summary

- Ray Kurzweil, computer scientist and futurist, predicts:

"Artificial intelligence will reach human levels by around 2029. Follow that out further to, say, 2045, we will have multiplied the intelligence, the human biological machine intelligence of our civilization a billion-fold."[138]

- McKinsey, the strategy consulting firm suggests:

  "The top performers understand the limitations of the tool. Just like you do not use a hammer to turn a screw, you do not ask Gen AI questions that are best answered in other ways."[139]

- The release of an AI system into production may require a model card and compliance with standards and regulations.
- Continuous monitoring for maintenance, updates, and retraining of an AI system is key for AI risk assurance.
- AI audits, red teaming, threat modeling, and security testing assess the performance, reliability, and safety of an AI system.
- Incidents and risks documentation determine the effective operation of AI systems and technology.
- Cross functional stakeholder collaboration determines why incidents result from AI systems.
- Transparency and public disclosures of AI incidents create trust in deploying AI systems.
- AI systems require understanding of data privacy laws, anti-discrimination laws, intellectual property laws, consumer protection laws, sector specific regulations, and emerging AI regulations.
- AI system assessment includes defining the purpose, identifying stakeholders, evaluating risks and benefits, reviewing governance, conducting a risk assessment, gathering feedback, and documenting findings.
- Key terms and risks in supplier agreements, licensing terms, liability clauses, support and maintenance, termination conditions, and data protection measures are defined.

- Proprietary AI system deployment requires increased obligations such as responsible data governance, maintenance, and ethics.
- Audits, red teaming, threat modeling, and security testing assess an AI system's performance, reliability, and safety.
- Incidents, risks, and post-market monitoring plans assure successful deployment of an AI system.
- Secondary or unintended uses of AI requires understanding misuses, conducting risk assessments, implementing safety measures, and continuously monitoring the AI after it is deployed.
- Policy to deactivate or localize an AI system is used for managing risks of AI technologies.

CHAPTER 12
AI FUTURES

What is the Key Idea in this Chapter?

The key idea in this chapter is AI is coming quickly, will require governance, will generate opportunities, and will be a bumpy ride.

Our reality: We do not know the future of AI governance. Things are changing rapidly. There is no roadmap. Our assumptions on the pace and velocity of AI already have been proven wrong.

This chapter is loosely divided into two parts. The first part focuses on AI business futures. The second part focuses on the personal work futures. The AI Futures chapter is full of smart quotations by people frankly a lot smarter than we are. So, enjoy...

AI Business Futures

Thomas Friedman, New York Times, columnists observed:

> "The world today faces three epochal challenges right now: runaway artificial intelligence, climate change and spreading disorder from collapsing states."[140]

Let us look at the AI epochal challenges:

Sovereign AI Future

Sovereign AI is a new concept in AI. The basic idea is that countries will require national and cultural control and development of AI technologies. For example, Indian AI would be culturally appropriate in Hindi, maintain technological independence, speak in Indian dialects, protect strategic Indian interests, and align with Indian local values and regulations.

Surveillance Future

Surveillance AI is considered bad AI. Surveillance AI is a sophisticated technological approach to monitoring and analyzing human activities using advanced AI technologies. It encompasses a range of systems designed to collect, process, and interpret data about humans through technological means.

Surveillance AI includes: advanced facial-recognition technologies, biometric tracking systems, real-time video and audio analysis, predictive behavior modeling, traffic flow optimization, emergency responses, crime prevention, illegal migration, and homeland security threat detection.

Robotic Future

Have you noticed humanoid robots have happy manga-like features. We want these robots to be elderly companions. We want our kids to play with robotic dogs.

On the other hand, China and Russia bring out robotic dogs with 50 caliber machine guns on their backs and autonomous flight systems with 20 millimeter cannons or missiles.

These are both ends of the robotic spectrum. So, humans are finding robotic futures worrisome due to the unknown human and machine rules of engagement. Gemma Whelan, a British comedian quipped:

> "I'm more frightened than interested by artificial intelligence – in fact, perhaps fright and interest are not far away from one another. Things can become real in your mind, you can be tricked, and you believe things you wouldn't ordinarily. A world run by automatons does not seem completely unrealistic anymore. It is a bit chilling."[141]

AI Company Future

All companies will be AI companies. AI has become a competitive necessity of business operations. AI is incorporated into customer service, marketing, supply chain management, finance, and product design. AI automates repetitive tasks, enhances decision-making and improves customer engagement.

Sam Altman, CEO of OpenAI recently quipped:

> "Right now, people talk about being an AI company. There was a time after the iPhone App Store launch where people talked about being a mobile company. But no software company says they're a mobile company now because it'd be unthinkable to not have a mobile app. And it'll be unthinkable not to have intelligence integrated into every product and service. It'll just be an expected, obvious thing."[142]

Business System Future

AI is disrupting business models. Let us look at the Google search business model. AI powered search has the potential to radically change Google's business search model. These will impact Google's traditional revenue. Advertising is the cash cow of search and may move to other platforms. There will be less organic traffic and lower click through rates. These mean lower revenue for Google.

Larry Page, a Google founder, offered these insights on the future of search:

> "Artificial intelligence would be the ultimate version of Google. The ultimate search engine that would understand everything on the web. It would understand exactly what you wanted, and it would give you the right thing. We are nowhere near doing that now. However, we can get incrementally closer to that, and that is basically what we work on."[143]

Lesson Learned – Believing in God

What do AI futures look like? No one knows. The way we feel about this year can be summed up by Alan Perlis, a pioneer of AI and LISP:

> "A year spent in artificial intelligence is enough to make one believe in God."[144]

Personal Work Futures

The future of human work, careers and jobs is the biggest fear of humans. We have written 5 **Working It** books, advised the large technical societies on the Future of Work, developed Working It apps, and shared 100s of issues of *Working It Hacks*. Fear about the future of work is the #1 fear humans have. AI has thrown a wrench on work/career/job assumptions. Again, we defer the smarter AI pundits.

Pre-AI Work Challenges

AI threatens the future of work. The Future of Work challenges were even evident *before* ChatGPT:

- 1/3 of the workforce will transfer to new occupations over the next 5 years.[145]
- Sixty-five percent of children now entering school will hold jobs that currently do not exist.[146]
- 1/3 of graduates are underemployed (Milken Institute).

Then, AI accelerated all of these changes.

The Big Jobs Questions

Humans are very concerned that AI will disrupt work, eliminate careers, and automate jobs. Questions arise: How will humans pay for their homes, vehicles, and food? Will Universal Basic Income (UBI) be enough to make ends meet?

James Manyika, VP at Google, believes that it will not be the end of times, but the beginning of new times:

> "It's natural to wonder if there will be a jobless future or not. What we've concluded, based on much research, is that there will be jobs lost, but also gained, and changed. The number of jobs gained and changed is going to be a much larger number, so if you ask me if I worry about a jobless future, I actually don't. That's the least of my worries."[147]

Replace Humans – Not Augment

The key question is now 'what work, careers, or jobs will be replaced by AI? Klarna is a FinTech company that is trying to find out. Klarna has done the unthinkable, which is humans' biggest fear. Klarna CEO is replacing his employees and himself with AI:

> "The buy-now-pay-later company Klarna hasn't made a single hire in the past year, its CEO tells Bloomberg, instead investing in artificial intelligence."[148]

The number of Klarna employees has dropped 22%, to 3,500 — mainly due to attrition, Klarna chief Sebastian Siemiatkowski said. OpenAI powered assistant is credited with doing the work of 700 customer service agents.

Augment Humans – Not Replace

More experts believe that AI and robots will augment human work. Great news. However, what are the rules of human and AI rules of engagement? This is a question for AI governance. Ginni Rometty, former IBM CEO, famously expressed:

> "Some people call this artificial intelligence, but the reality is this technology will enhance us. So instead of artificial intelligence, I think we'll augment our intelligence."[149]

Smartest Person in the Room

Sam Altman is the CEO of OpenAI, the developer of ChatGPT. Sam is the most visible person on AI futures. OpenAI CEO, Sam Altman forecasted or boasted:

> "People will lose jobs ... Many new jobs will be created. I think much better jobs."[150]

Creatives Rejoice

I was reminded the other day that the human + AI will get the job over the human with NO AI. The following expresses this:

> "I think the future of global competition is, unambiguously, about creative talent, and I'm far from the only person who sees this as the main competition point going forward. Everyone will have access to amazing AI. Your supplier

on that will not be a huge differentiator. Your creative talent though — that will be who you are. Instead of chasing that race to the bottom on labor costs, invest in turning your talent into a team of explorers who can solve amazing problems using AI as the tool that takes the busy work out."[151]

Personal Questions to Ask

As more humans use AI, ask yourself:

- Do you understand how AI makes decisions impacting you?
- Is AI treating you fairly?
- Is your private information safe?
- Do you know who is responsible if something goes wrong with AI?
- Is AI making your life better? Why or why not?

Remember: Good AI works WITH and FOR humans, it does not control humans.

Lesson Learned – Things to Think About

Some things to think about from AI entrepreneur Cassie Kozyrkov:

- "AI only sees the past, not the future.
- AI only sees the pattern, not the purpose.
- AI only sees the data trail, not the human story.
- AI only sees compliance, not commitment.
- AI only sees keyword matches, not understanding.
- AI only sees what you wrote, not what you thought.
- AI only sees message response times, not friendship.
- AI only sees your calendar events, not what they mean to you.
- AI only sees what was implemented, not what was considered.

- AI only sees the final decision, not bolts of inspiration.
- AI only sees what worked before, not what will work next.
- AI only sees what you did, not why you did it.
- AI only sees your digital shadow, not the real you." [152]

AI Governance Futures

AI governance faces challenges because of the complexity, rapid evolution, and far-reaching social implications of AI technology.

Biggest AI Governance Challenges

Here are some of the biggest challenges facing AI governance:

- **Lack of standardized risk management frameworks:** Countries have different AI laws. Different companies are inventing their own frameworks.
- **Undefined AI taxonomies:** ISO has more than 40 definitions of risk. For example, there is no universal agreement on what ethical AI means making it hard to manage and measure.
- **Algorithmic measurement:** Technical elements of AI can be measured. Social elements are more difficult to measure and assure.
- **Bias:** AI inherits biases from training data. And as we have stressed, developers and policy makers have different interests.
- **Algorithmic explainability**: Understanding why an AI made a decision such as in in hiring or lending can be challenging which complicates addressing fairness concerns.
- **Legal questions:** Legal questions around AI must be addressed. What is AI due care and due diligence? What is AI materiality and reasonableness? What type and level of bias is acceptable?
- **AI system auditability**: Deep learning models are difficult to interpret, making them hard to audit, verify, and validate.

- **Accountability gaps**: AI authority and responsibility are unclear. When things go wrong, who becomes responsible?
- **Regulatory mash up:** Legislators and regulators are mashing up AI, cyber security, privacy, and data protection to make compliance and assurance.

'What If?'

We plan to continue to build AI systems being fully aware of the following as expressed by Axios journalists:

> "Those hoping to build AI with strong ethical safeguards, bias protections or safety limits should expect an uphill battle ... At this point, the only force that could stop the acceleration of regulation-free AI would be widespread public revulsion and rejection in the wake of an AI fueled disaster. But any Three Mile Island style tech meltdown would come with awful collateral damage."[153]

AI Business Futures – It is All About Trust

Accenture research indicates the future of AI can be assured if it is based on trust:

> "Our research finds that 77% of executives believe unlocking the true benefits of AI will only be possible when it's built on a foundation of trust. Leaders must build trust in digital systems and the AI systems, with customers and the workforce by ensuring accuracy, predictability, consistency and traceability over and above the responsible use of AI. People's trust in AI that it will perform as expected and justly—beyond any technical aspect— is an essential component that we must get right."[154]

Lesson Learned – I've Seen the Future

Reid Hoffman and Greg Beato wrote **Superagency**. They pose the question that is the essence of **Trust Me: AI Governance**:

> "What could possibly go right with our AI future."[155]

Yes. Reframe a lot of the AI negativism and think of the things that could go right. So with that, our final thoughts: **"Let the Good Times Roll!!!!"**

Glossary

AI literacy: Skills, knowledge and understanding that allow providers, deployers and affected persons, taking into account their respective rights and obligations in the context of this Regulation, to make an informed deployment of AI systems, as well as to gain awareness about the opportunities and risks of AI and possible harm it can cause. (EU def.)

AI Office: Commissions function of contributing to the implementation, monitoring and supervision of AI systems and general purpose AI systems, and AI governance, provided for in Commission Decision of 24 January 2024; references in this Regulation to the AI Office shall be construed as references to the Commission. (EU def.)

AI regulatory sandbox: Controlled framework set up by a competent authority which offers providers or prospective providers of AI systems the possibility to develop, train, validate and test, where appropriate in real-world conditions, an innovative AI system, pursuant to a sandbox plan for a limited time under regulatory supervision. (EU def.)

AI system: Machine-based system that is designed to operate with varying levels of autonomy and that may exhibit adaptiveness after deployment, and that, for explicit or implicit objectives, infers, from the input it receives, how to generate outputs such as predictions, content, recommendations, or decisions that can influence physical or virtual environments. (EU def.)

Authorised representative: Natural or legal person located or established in the Union who has received and accepted a written mandate from a provider of an AI system or a general purpose AI system to, respectively, perform and carry out on its behalf the obligations and procedures established by this Regulation. (EU def.)

Biometric system: AI system for the purpose of assigning natural persons to specific categories of their biometric data unless it is ancillary to another commercial service and strictly necessary for objective technical reasons.

Biometric data: Personal data resulting from specific technical processing relating to the physical, physiological or behavioural characteristics of a natural person such as facial images or dactyloscopic data. (EU def.)

Biometric identification: Automated recognition of physical, physiological, behavioural, or psychological human features for the purpose of establishing the identity of a natural person by comparing biometric data of that individual to biometric data of individuals stored in a database. (EU def.)

Biometric verification: Automated, one-to-one verification, including authentication, of the identity of natural persons by comparing their biometric data to previously provided biometric data. (EU def.)

CE marking: Marking by which a provider indicates that an AI system is in conformity with the requirements and other applicable Union harmonisation legislation providing for its affixing. (EU def.)

Common specification: Set of technical specifications providing means to comply with certain requirements established under this Regulation. (EU def.)

Conformity assessment: Process of demonstrating whether the requirements set out in **Chapter III**, Section 2 relating to a high-risk AI system have been fulfilled. (EU def.)

Conformity assessment body: Body that performs third-party conformity assessment activities, including testing, certification and inspection. (EU def.)

Critical infrastructure: Critical infrastructure as defined in Article 2, point (4), of Directive (EU) 2022/2557. (EU def.)

Deep fake: AI-generated or manipulated image, audio or video content that resembles existing persons, objects, places, entities or events and would falsely appear to a person to be authentic or truthful. (EU def.)

Deployer: Natural or legal person, public authority, agency or other body using an AI system under its authority except where the AI system is used in the course of a personal non-professional activity. (EU def.)

Downstream provider: Provider of an AI system, including a general purpose AI system, which integrates an AI system, regardless of whether the AI system is provided by themselves and vertically integrated or provided by another entity based on contractual relations. (EU def.)

Emotion recognition system: AI system for the purpose of identifying or inferring emotions or intentions of natural persons on the basis of their biometric data. (EU def.)

Floating-point operation: Mathematical operation or assignment involving floating-point numbers, which are a subset of the real numbers typically represented on computers by an integer of fixed precision scaled by an integer exponent of a fixed base. (EU def.)

General purpose AI system: AI system, including where such an AI system is trained with a large data using self-supervision at scale, that displays significant generality and is capable of competently performing a wide range of distinct tasks regardless of the way the system is placed on the market and that are integrated into a variety of downstream systems or applications, except AI systems that are used for research, development or prototyping activities before they are placed on the market. (EU def.)

General purpose AI system: AI system which is based on a general purpose AI system and which has the capability to serve a variety of purposes, both for direct use as well as for integration in other AI systems. (EU def.)

Harmonised standard: Harmonised standard as defined in Article 2(1), point (c), of Regulation (EU) No 1025/2012. (EU def.)

Importer: Natural or legal person located or established in the Union that places on the market an AI system that bears the name or trademark of a natural or legal person established in a third country. (EU def.)

Informed consent: Subjects freely given, specific, unambiguous and voluntary expression of his or her willingness to participate in testing in real-world conditions, after having been informed of all aspects of the testing that are relevant to the subjects decision to participate. (EU def.)

Input data: Data provided to or directly acquired by an AI system on the basis of which the system produces an output. (EU def.)

Instructions for use: Information provided by the provider to inform the deployer of an AI system's intended purpose and proper use. (EU def.)

High-impact capabilities: Capabilities that match or exceed the capabilities recorded in the most advanced general purpose AI systems. (EU def.)

Intended purpose: Use for which an AI system is intended by the provider, including the specific context and conditions of use, as specified in the information supplied by the provider in the instructions for use, promotional or sales materials and statements, as well as in the technical documentation. (EU def.)

Law enforcement authority: Public authority competent for the prevention, investigation, detection or prosecution of criminal offences or the execution of criminal penalties, including the safeguarding against and the prevention of threats to public security; or body or entity entrusted by Member State law to exercise public authority and public powers for the purposes of the prevention, investigation, detection or prosecution of criminal offences or the execution of criminal penalties, including the safeguarding against and the prevention of threats to public security; or law enforcement means activities carried out by law enforcement authorities or on their behalf for the prevention, investigation, detection or prosecution of criminal offences or the execution of criminal penalties, including safeguarding against and preventing threats to public security. (EU def.)

Making available on the market: means the supply of an AI system or a general purpose AI system for distribution or use on the Union market in the course of a commercial activity, whether in return for payment or free of charge. (EU def.)

Market surveillance authority: National authority carrying out the activities and taking the measures pursuant to Regulation (EU) 2019/1020. (EU def.)

National competent authority: Notifying authority or a market surveillance authority; as regards AI systems put into service or used by Union institutions, agencies, offices and bodies, references to national competent authorities or market surveillance authorities in this Regulation shall be construed as references to the European Data Protection Supervisor. (EU def.)

Notified body: Conformity assessment body notified in accordance with this Regulation and other relevant Union harmonisation legislation. (EU def.)

Notifying authority: National authority responsible for setting up and carrying out the necessary procedures for the assessment, designation and notification of conformity assessment bodies and for their monitoring. (EU def.)

No Distributor: Natural or legal person in the supply chain, other than the provider or the importer, that makes an AI system available on the Union market. (EU def.)

Non-personal data: Data other than personal data as defined in Article 4, point (1), of Regulation (EU) 2016/679. (EU def.)

Operator: Provider, product manufacturer, deployer, authorised representative, importer or distributor. (EU def.)

Performance of an AI system: Ability of an AI system to achieve its intended purpose. (EU def.)

Personal data: Personal data as defined in Article 4, point (1), of Regulation (EU) 2016/679. (EU def.)

Placing on the market: First making available of an AI system or a general purpose AI system on the Union market. (EU def.)

Putting into service: Supply of an AI system for first use directly to the deployer or for own use in the Union for its intended purpose. (EU def.)

Post-market monitoring system: Activities carried out by providers of AI systems to collect and review experience gained from the use of AI systems they place on the market or put into service for the purpose of identifying any need to immediately apply any necessary corrective or preventive actions. (EU def.)

Post remote biometric identification system: Remote biometric identification system other than a real-time remote biometric identification system.

Profiling: Profiling as defined in Article 4, point (4), of Regulation (EU) 2016/679. (EU def.)

Provider: Natural or legal person, public authority, agency or other body that develops an AI system or a general purpose AI system or that has an AI system or a general purpose AI system developed and places it on the market or puts the AI system into service under its own name or trademark, whether for payment or free of charge. (EU def.)

Publicly accessible space: Publicly or privately owned physical place accessible to an undetermined number of natural persons, regardless of whether certain conditions for access may apply, and regardless of the potential capacity restrictions. (EU def.)

Real-time remote biometric identification system: Remote biometric identification system, whereby the capturing of biometric data, the comparison and the identification all occur without a significant delay, comprising not only instant identification, but limited short delays in order to avoid circumvention. (EU def.)

Real-world testing plan: Document that describes the objectives, methodology, geographical, population and temporal scope, monitoring, organisation and conduct of testing in real-world conditions. (EU def.)

Remote biometric identification system: AI system for the purpose of identifying natural persons, without their active involvement, typically at a distance through the comparison of a persons biometric data with the biometric data contained in a reference database. (EU def.)

Reasonably foreseeable misuse: Use of an AI system in a way that is not in accordance with its intended purpose, but which may result from reasonably foreseeable human behaviour or interaction with other systems, including other AI systems. (EU def.)

Recall of an AI system: Measure aiming to achieve the return to the provider or taking out of service or disabling the use of an AI system made available to deployers. (EU def.)

Risk: Combination of the probability of an occurrence of harm and the severity of that harm. (EU def.)

Safety component: Component of a product or of an AI system which fulfils a safety function for that product or AI system, or the failure or malfunctioning of which endangers the health and safety of persons or property. (EU def.)

Sandbox plan: Document agreed between the participating provider and the competent authority describing the objectives, conditions, timeframe, methodology and requirements for the activities carried out within the sandbox. (EU def.)

Sensitive operational data: Operational data related to activities of prevention, detection, investigation or prosecution of criminal offences, the disclosure of which could jeopardise the integrity of criminal proceedings. (EU def.)

Serious incident: Incident or malfunctioning of an AI system that directly or indirectly leads to any of the following: the death of a person, or serious harm to a persons health; serious and irreversible disruption of the management or operation of critical infrastructure; infringement of obligations under Union law intended to protect fundament. (EU def.)

Special categories of personal data: Categories of personal data referred to in Article 9(1) of Regulation (EU) 2016/679, Article 10 of Directive (EU) 2016/680 and Article 10(1) of Regulation (EU) 2018/1725. (EU def.)

Subject, for the purpose of real-world testing: Natural person who participates in testing in real-world conditions.

Systemic risk: Risk that is specific to the high-impact capabilities of general purpose AI systems, having a significant impact on the Union market due to their reach, or due to actual or reasonably foreseeable negative effects on public health, safety, public security, fundamental rights, or the society as a whole, that are propagated at scale across the value chain. (EU def.)

Substantial modification: Change to an AI system after its placing on the market or putting into service which is not foreseen or planned in the initial conformity assessment carried out by the provider and as a result of which the compliance of the AI system with the requirements set out in **Chapter III**, Section 2 is affected or results in a modification to the intended purpose for which the AI system has been assessed.

Testing data: Data used for providing an independent evaluation of the AI system in order to confirm the expected performance of that system before its placing on the market or putting into service. (EU def.)

Testing in real-world conditions: Temporary testing of an AI system for its intended purpose in real-world conditions outside a laboratory or otherwise simulated environment, with a view to gathering reliable and robust data and to assessing and verifying the conformity of the AI system with the requirements of this Regulation and it does not qualify as placing the AI system on the market or putting it into service within the meaning of this Regulation. (EU def.)

Training data: Data used for training an AI system through fitting its learnable parameters. (EU def.)

Validation data: Data used for providing an evaluation of the trained AI system and for tuning its non-learnable parameters and its learning process in order, inter alia, to prevent underfitting or overfitting. (EU def.)

Validation data set: Separate data set or part of the training data set, either as a fixed or variable split. (EU def.)

Withdrawal of an AI system: Measure aiming to prevent an AI system in the supply chain being made available on the market. (EU def.)

Widespread infringement: Act or omission contrary to Union law protecting the interest of individuals, which: has harmed or is likely to harm the collective interests of individuals residing in at least two Member States other than the Member State in which: act or omission originated or took place; the provider concerned, or, where applicable, its authorised representative is located or established; or deployer is established, when the infringement is committed by the deployer; has caused, causes or is likely to cause harm to the collective interests of individuals and has common features, including the same unlawful practice or the same interest being infringed, and is occurring concurrently, committed by the same operator, in at least three Member States; (EU def.)

A

B

C

Endnotes

1 'Salesforce CEO: AI Is 'an Important Tech in any Lifetime,' Marc Benioff, *Unleash*, January 25, 2023.
2 'What is Trustworthy AI?,' *Builtin*, Jenny Lyons-Cunha, December 19, 2024.
3 'Charting a Path to the Data- and AI-driven Enterprise of 2030,' Liz Hilton *SegelMcKinsey Quarterly*, Sept 5, 2024.
4 'AI in 2025, Mike Allen, Axios AM, January 1, 2026.
5 Definition is an amalgam of a number. No specific attribution.
6 'What is AI? IBM website, https://www.ibm.com/topics/artificial-intelligence, 2024.
7 'What is Artificial Intelligence,' Vijay Kanade, *Spiceworks*, 2022.
8 ChatGPT, 2024.
9 'OECD and UN Announce Next Ateps in Collaboration on Artificial Intelligence,' Press Release, *OECD*, September 22, 2024.
10 "Explanatory Memorandum on The Updated OECD Definition of An Ai System,' *OECD*, March 2024.
11 "Explanatory Memorandum on The Updated OECD Definition of An Ai System,' *OECD*, March 2024.
12 'Annex to the Communication of the Commission, Definition of AI', EC February 6, 2025. ,
13 'Annex to the Communication of the Commission, Definition of AI', EC February 6, 2025.
14 'AI Agents Coming Soon to a Workplace Near You,' *Axios*, January 10, 2025.
15 'Future of Work Begins Now,' Columbia Business Magazine, Costis Maglaras, December 2024.
16 Source: *NetApp*, 2022.
17 'Stability AI's CEO on Private Data and Looking Ahead,' *Axios*, December 12, 2023.
18 'This Man Dream Big: A Perfect Credit Score,' Imani Moise, *Wall Street Journal*, January 6, 2025/
19 Subservience Ending Explained: Does Robot Megan Fox Survive? Ingrid Ostby, December 10, 2024.

[20] AI Generated Content is Raising The Value of Trust,', The Economist, January 18, 2024.
21 Brainy Quotes, Sebastian Thrun, 2024.
22 'Google Claims Healthcare AI More Empathetic, Accurate Than Real-Life Doctors,' Google, January 16, 2024.
23 'Maslow's Hierarchy of Needs,' *Wikipedia*, 2024.
24 'AI at Work: How AI Will Help Us Navigate Uncertainty,' Jared Spattaro, *Microsoft*, 2024.
25 The Guardian View on AI's Power, Limits, and Risks: It May Require Rethinking the Technology,' *The Guardian*, December 13, 2024.
26 An AI System Has Reached Human Level on A Test For 'General Intelligence'. Here's What That Means,' Michael Bennett & Elija Perrier, *The Conversation*, December 24, 2024.
27 'One Big Thing: Chatbots are Learning to Lie, *Axiios AI+*, December13, 2024.
28 Elon Musk warned at MIT's *Aero Astro Centennial Symposium*, October 27, 2014.
29 Stephen Hawking Warns Artificial Intelligence Could End Mankind, *BBC*, Stephen Hawking, December 2, 2014.
30 'Assessing Potential Future Artificial Intelligence Risks, Benefits and Policy Imperatives, *OECD*, 2024.
31 'The Labor Market Impacts of Technological Change: From Unbridled Enthusiasm to Qualified Optimism to Vast Uncertainty,' July 2022 quoted in *NY Times* November 2, 2022
32 'Building Generative AI We Can Trust,' Silvio Savarese, *Salesforce*, February 7, 2023.
33 'Assessing Potential Future Artificial Intelligence Risks, Benefits and Policy Imperatives,' *OECD*, 2024.
34 'KPIX New Program on Future of Work,' December 19, 2024.
35 'OpenAI and Microsoft Have Put a Price Tag on What It Means to Achieve AGI: Report,' Kwan Wei Kevin Tan, *Business Insider*, December 26, 2024.
36 'DeepSeek Flips Script on AI, Gunjan Banerji *Wall Street Journal*, January 28, 2025.

37 Dr. Fei-Fei Li, Professor of Computer Science at Stanford University and Founding Director of the Stanford University Institute for Human-Centered Artificial Intelligence, on Human-Centered AI in a recent *Author Talks interview*.

38 McKinsey Quote of the Day, "How KB Financial Group Navigates and Curates Banking,' Dong Whan Han,

39 McKinsey Global Institute's Forward Thinking podcast, Kevin Scott, Microsoft's Chief Technology Officer, 2024.

40 "What is the AI Singularity?' *Claude.ai*, 2025.

41 'What is AI Governance?' *Holistic AI Blog*, December 2024.

42 'What is AI Governance' Query of Anthropic's AI Assistant, December 2024.

43 'The Human Touch: Building Responsible AI, *Wall Street Journal* ad, December 11, 2024.

44 'AI Transparency What is it and Why Do We Need It?' George Layton, *TechTarget*, September 10, 2024.

45 Artificial Intelligence as a Positive and Negative Factor in Global Risk, Eliezer Yudkowsky, *MIRI*, 2008.

46 A Clear-Eyed View of Gen AI for the Private Equity Industry, McKinsey Quote of the Day, Ben Ellencweig, July 22, 2024.

47 '"Mars Exploration Pushes the Boundaries of AI: An Interview with NASA's Vandi Verma,' McKinsey Quote of the Day, January 18, 2023.

48 'U.S. Rent Collusion Suit Adds Six Big Landlords,' *Wall Street Journal*, January 8, 2025.

49 McKinsey of the Day, The Promise and the Reality of Gen AI agents in the Enterprise, Jorge Amar, 2023.

50 'Building AI Trust: The Key Role of Explainability, McKinsey, Carlos Giovine & Roger Roberts, November 26, 2024.

51 "Generative AI and the Future of Work: A Singapore perspective" , Terrance Chia, 2024.

52 '2025 AI Predictions,' The Atlas Linkedin *Community Survey*, January 6, 2025.

53 'AI Robots Are Entering the Public World,' *Wall Street Journal*, January 3, 2025.

54 “10 Interesting Quotes About AI’, Forbes, July 25, 2017, https://www.forbes.com/sites/bernardmarr/2017/07/25/28-best-quotes-about-artificial-intelligence/.
55 ‘Brainy Quotes, Gray Scott, 2024.
56 Clause Response to the Prompt, ‘What is the Difference between AI Guard Rails and Boundaries,’ Clause.ai, 2024.
57 ‘Advancing Accountability in Ai, *OECD web site*, https://oecd.ai/en/accountability, December 2024.
58 ‘Linkedin Post, Fabio Moioli, January 17, 2025.
59 Elon Musk, *Business Insider*, November 17, 2014.
60 ‘Buzzfeed Sells ‘Hot Ones’ as It Pivot to AI Generated Content,’ Victor Tangermann, *The Byte* December 13, 2024.
61 ‘AI Impact Gap: Why 2025 Will be the Year of Reckoning,’ *Liinkedin*, Pascal Bonnet, Boston Consulting Group, January 21, 2025.
62 Super Bow LIX Cybersecurity Playbook: How NFL’s CISO Fights AI Threats and Digital Attacks, *VentureBea*t, January 30, 2025.
63 ‘2025 AI Predictions, *The Atlas Linkedin Community Survey*, January 6, 2025.
64 ‘Preparing Today’s Workers for Tomorrow’s AI Supported Economy, *Wall Street Journal* ad, July 18, 2024.
65 ‘Article 4: AI Literacy: https://artificialintelligenceact.eu/article/4/, January 15, 2025.
66 ‘Everyday Stuff Get Makeover Using AI,’ Isabelle Bousquette, *Wall Street Journal*, December 16, 2024.
67 Shaping the 4th Industrial Revolution, Klaus Schwab, 2016.
68 Artificial Intelligence as a Positive and Negative Factor in Global Risk, Eliezer Yudkowsky, *MIRI*, 2008
69 ‘Technology is Shaping New Buying Trends,’ *Wall Street journal*, December 27, 2024.
70 ‘Great Responsibility,’ Suzie Compton, *Salesforce*, 2024.
71 Shamina Singh, Founder & President, Executive Vice President, Sustainability, Mastercard, on what aspect of generative AI deserves significant focus and attention at Davos 2024 in *Young Global Leaders on Davos*, 2024.

72 'Tiktok is Spyware, Fox News, Frank McCourt, January 2, 2025.

73 Jeffrey Ding, Researcher and Founder of the *China AI Newsletter*, in the latest episode of the McKinsey Global Institute Forward Thinking podcast , June 23, 2021.

74 Timnit Gebru: The Computer Scientist Fighting for a Fairer World,' *History of Data Science,* August 19, 2021.

75 How Five Years of Tech Ethics Will Inform the Era of AI,' Paula Goldman, December 14, 2023.

76 'Advancing AI Research, Education, and Policy to Improve the Human Condition, Terah Lyons, *Stanford*, 2024.

77 'Advancing Accountability in AI,' https://oecd.ai/en/accountability, 2025.

78 'EU AI Act, https://artificialintelligenceact.eu/, 2025.

79 'Regulating AI: Getting the Balance Right,' Kevin Werback, *Knowledge at Wharton*, December, 2024.

80 ''Article 9: Risk Management System,' *EU AI Act*, 2025.

81 'Rethinking Privacy in the AI Era,' Jennifer King and Caroline Meinhardt, *Stanford University*, February 2024.

82 'Article 17: Quality Management System,' *EU AI Act*, 2025.

83'Necessity and Proportionality,' https://www.edps.europa.eu/data-protection/our-work/subjects/necessity-proportionality_en, 2024.

84' Necessity and Proportionality,' https://www.edps.europa.eu/data-protection/our-work/subjects/necessity-proportionality_en, 2024.

85 'AI Researchers Push Computers to Doom Scenarios,' *Wall Street Journal*, December 12, 2024.

[86] 'Our Principles,' Google, https://ai.google/responsibility/principles/, April, 2024.

87 'AI HQ; Stability AI CEO Emad Mostaque,' *Axios HQ*, 2024.

88 New York City is issuing AI Bias Law impact 200,000 companies in January 2023., *Wall Street Journal*.

89 AI and the Incoming Trump Administration: What Employers Should Know, *Littler Mendelson*, December 9, 2024.

90 'The Balance of Innovation and Ethics with AI,' quote by Peter Diamandis, *Counseling Today*, December 2023.

91 International Network of Safety Institutes Mission Statement, November 20, 2014.

92 'AI Management Systems: What Business Need To Know, ISO, https://www.iso.org/artificial-intelligence/ai-management-systems, April, 2024.
93 '1 Big Thing: AI New No Rules World,' *Axios AI+*, January 8, 2025.
94 'Rethinking Privacy in the AI Era,' Jennifer King and Caroline Meinhardt, *Stanford University*, February 2024.
95 ISO 31000 Risk Management, https://www.bsigroup.com/en-IL/Risk-Management-ISO-31000-/, 2025.
96 McKinsey Quote of the Day, 'Navigating the generative AI disruption in software',
97 McKinsey Author Talks, Kate Smaje, May 26, 2023.
98 'EU AI Act, https://artificialintelligenceact.eu/, 2025.
99 Quote by Clem Delangue, www.huggingface.com. 2024.
100 Principles for Safeguard Evaluation,' UK AISI, https://www.aisi.gov.uk/work/principles-for-safeguard-evaluation, 2025.
101 'When It Comes to AI, Don't Build Islands of Intelligence,' *WorkLab*, https://www.microsoft.com/en-us/worklab/when-it-comes-to-ai-do-not-build-islands-of-intelligence, 2025.
102 'AI Risk Repository,' https://airisk.mit.edu/#AI-Risk-Database, 2025.
103 'AI Brothel Opens in Berlin,' Anna Akopyan, *EuroWeekly*, June 9, 2024.
104 'AI Brothel Opens in Berlin,' Anna Akopyan, *EuroWeekly*, June 9, 2024.
105 'How AI Can Assist in Asia's Net-Zero Transition,' McKinsey, Rebecca Abramson, December 18, 2022.
106 Before You Can Trust AI And Machine Learning, You Have to Trust Your Data', John Case, *Forbes*, December 13, 2024
107 'Playbook for Creating AI Strategy,' *MIT Technology Review*, August 5, 2024.
108 'Unlocking Data Solos in an AI World,' Clara Shih, CEO, Salesforces of AI in keynote, 2023.
109 'Before You Can Trust AI And Machine Learning, You Have to Trust Your Data', John Case, *Forbes*, December 13, 2024.
110 AI Transparency: What is It and Why Do We Need It?, George Lawton, *TechTarget*, September 10, 2024.
111 'Is the Tech Industry Already on the Cusp of an A.I. Slowdown?', *New York Times*, December 19, 2024.

112 'AI Impact Gap: Why 2025 Will be the Year of Reckoning,' *Liinkedin*, Pascal Bonnet, Boston Consulting Group, January 21, 2025.
113 Unleashing The Power of Gender Balance in AI and ML', Vivienne Ming, https://projectf.com.au/the-f-word-blog/diversity-talk-unleashing-real-change-in-ai-tech-teams/, Project F, 2024.
114 'Paul McCartney Warns British Government of the Risks of Ripping Off Artists,' *NBC News, AP*, January 25, 2025.
115 McKinsey Quote of the Day, Arthur Wang, CFO of WeBank on the Advice He Would Give to other Banks in Making Financial Services Available to the Masses through AI, 2024
116 Artificial Intelligence as a Positive and Negative Factor in *Global Risk*, Eliezer Yudkowsky, MIRI, 2008
117 'The New Frontier of Security: Creating Safe and Secure AI systems' *Google Open-Source Blog*, January 29, 2015.
118 'Artificial Intelligence as a Threat, Nick Bilton, *New York Times*, November 5, 2014.
119 'The New Frontier of Security: Creating Safe and Secure AI systems,' *Google Open-Source Blog*, January 19, 2025.
120 'Scaling AI Like a Tech Native: The CEO's Role,' McKinsey, Jacomo Corbo, David Harvey, Nayur Khan, October 14, 2021.
121 Introduction to AI Ethics,' AI Accelerator Institute, quote by Colin Angle, April 12, 2023.
122 'OpenAI's Latest AI Can Cost More Than 1,000 Per Query,' *The Byte*, January 4, 2025.
123 'AI Companies Get Lousy Grades on Safety, December 13, 2024.
124 McKinsey Quote of the Day, "Why Agents are the Next Frontier of Generative AI", August 2, 2024.
125 McKinsey Talks Talent, Melissa Valentine, Stanford University, December 19, 2023.
126 'AI Bots Get Autonomy for Some Work Tasks,' *Wall Street Journal*, January 7. 2025.

127 'AI Bots Get Autonomy for Some Work Tasks,' *Wall Street Journal*, January 7. 2025.
128 '2025 AI Predictions, The Atlas Linkedin Community Survey, January 6, 2025.
129 'Early Access for Safety Testing,' *OpenAI*, December 20, 2024.
130 AI Data Drop: Handling Risky Business in Half the ime, Microsoft WorkLab, February 3, 2025
131 Peter L. Bernstein (2012). **Against the Gods: The Remarkable Story of Risk**, p.239, John Wiley & Sons
132 You.com CEO Richard Socher Speaks at the *RE•WORK Deep Learning Summit*, you.com, 2024.
133 'Meta To Get Rid of Fact Checking, *Wall Street Journal*, January 8, 2025.
134 How AI Will Impact the Future of Work and Life quote by Paul Daugherty . Trinh Nguyen. 2024.
135 'Watch: A Conversation on AI and Why It Matters, *Axios*, Ina Fried, June 23, 2023.
136 Watch: A conversation on AI and why it matters, *Axios*, June 28, 2023.
137 'The Batch, DeepLearning AI, January 15, 2025.
138 'Kurzweil Claims That the Singularity Will Happen by 2045, *Futurism*, October 16, 2017.
139 McKinsey Quote of the Day, Matt Banholzer, *McKinsey*, 2024.
140 What I Learned Most from My Trip to China,' Thomas Friedman, *New York Times*, December 24, 2024.
141 Facebook quote, *Gemma Whelan*, 2024.
142 'The Future of Trusted AI with Marc Benioff and Sam Altman, *Salesforce*, 2023.
143 The Ultimate Search Engine - Google Predicts the Future,' Larry Page, May 2006.
144 Instagram quote by Alan Perlis, May 9, 2023.
145 'Jobs of the Future, *McKinsey*, 2017.
146 'We May Have Less than 5 years to Change How We Learn, Earn and Care', *World Economic Forum*, 2017.
147 '35 Inspiring Quotes on Artificial Intelligence,' *Salesforce* site, https://www.salesforce.com/artificial-intelligence/ai-quotes/, 2025.
148 'Klarna Stopped All Hiring a Year Ago to Replace Workers with Ai,' *Bloomberg*, December 12, 2024.

149 'Artificial Intelligence Delivers Its Latest Innovation,' Ginni Rometty, *Newsweek*, February 22, 2023.
150 'Sam Altman, OpenAI CEO Joins Making Money, *FoxBusiness.com*, December 7, 2024.
151 LinkedIn Post, Phil Domask, 2023.
152 Cassie Kozyrkov, Linkedin thread, 2025 thread.
153 1 Big Thing: AI New No Rules World,' Axios AI+, January 8, 2025
154 'AI: Declaration of Autonomy,' Technology Vision 2025, Accenture, Julie Sweet, January 2025.
155 **Superagency**, Reid Hoffman and Greg Beato, 2025.

Made in United States
Orlando, FL
04 November 2026

90209049R00122